Main Currents in Sociological Thought

MAIN CURRENTS IN SOCIOLOGICAL THOUGHT

Volume I

Montesquieu, Comte, Marx, Tocqueville,
AND
The Sociologists and the Revolution of 1848

BY RAYMOND ARON

Translated from the French by
RICHARD HOWARD and HELEN WEAVER

ANCHOR BOOKS
DOUBLEDAY & COMPANY, INC.
GARDEN CITY, NEW YORK

Contents

Introduction

ANY HISTORICAL introduction to general sociology raises certain philosophical or methodological problems. Indeed, one cannot undertake the history of any historical entity—whether a nation or a scientific discipline—without establishing its definition, without fixing its limits. One cannot retrace the past of France until one has decided what France is. And if in this case the problem seems a simple one, substitute for the concept "France" the concept "Europe" or the concept of "the West," and you will soon see that the definition of a historical subject is not a matter of course.

To write a history of sociology, one must arrive at a definition of what is to be called "sociology." Now, in dealing with a natural science like mathematics or physics, the solution is relatively simple: in these cases, science presents itself in the form of a body of proven propositions; and history can, if so desired, be reduced to the gradual discovery of the truth as it appears to us today. In this case, history is no more than the discovery of currently accepted truths. But if we employed such a method in the case of sociology, the consequences would be unfortunate, for there is great danger that each sociologist would write a different history of sociology, each of such histories obligingly leading up to the current situation and the present truth, i.e., the truth held by the sociologist in question.

I cannot offer any body of demonstrated truths which might be called the present state of sociology. Strictly speaking, there is the present state of *French* sociology, illustrated in the *Traité de Sociologie* published by the

1

Presses Universitaires de France. But this treatise is not necessarily regarded as the last word on the subject in the United States, where a more conceptual, more analytic notion of sociology prevails; still less in the Soviet Union, where a more categorical notion of sociological truth is preferred.

To avoid this difficulty, I prefer to regard as sociology that which societies designate as such; I regard as sociologists those who assume this title—of honor or disgrace, as you will; and for an idea of sociology throughout the world, I refer to the quadrennial International Congress of Sociology attended by several hundred people who call themselves sociologists and who discuss a certain number of problems they call sociological—problems which imply a certain idea of what sociology is and should be today.

In the course of recent international congresses, two typical schools emerged, each aware of itself and of its opposition to the other. These two concepts of sociology were that of the American school on the one hand and that of the Soviet, or Marxist, school on the other. (To be sure, there appeared intermediate schools—the Polish school or the French school—which shared a tendency to cast aspersions on the two dominant ones: in these cases the sociologists reflected quite accurately the societies they represented.) I shall try to define briefly the respective characteristics of each of these two major schools of sociology.

Marxist sociology is essentially an inclusive interpretation of modern societies and of the evolution of social types. The primary object of sociological investigation, according to our colleagues in Moscow, is the discovery of the fundamental laws of historical evolution. These laws of historical evolution afford two major aspects, one static, the other dynamic. More precisely, the Soviet sociologists establish the fundamental structure of human societies as determined by the forces and relations of production; next, they define the principal social types and trace the necessary evolution from one social type to another. These social types are, moreover, the same ones discovered by Marx in the preface to his *Critique of Political Economy:* the ancient economy, based on slavery; the medieval economy, based on serfdom; the capitalist economy, based on wage earning; and finally, the socialist economy, which marks the end of the exploitation of man by man and of one class by another.

Only one of Marx's social types has virtually disappeared from contemporary Soviet literature, and that is the social type Marx called "the Asiatic mode of production," which has given rise to a great body of literature, both Marxist and non-Marxist. The Asiatic mode of production, as Marx defined it, was based on the dominant and almost exclusive role of the state in the organization of production. This role of the state was in turn related to the necessity of controlling the rivers and their irrigating functions. The Asiatic mode of production was not characterized by a class struggle within the society, but by a simple opposition between the mass of the ruled and the minority of the rulers or administrators organized into a vast state bureaucracy.

A sociology of the Marxist kind is synthetic, in the sense that Auguste Comte assigned to this term: it comprehends the whole of each society; unlike the specific social sciences, it is distinguished by its all-encompassing design. It seeks to grasp society in its totality, rather than any particular aspect of society.

This synthetic sociology is at the same time a historical sociology. It is not content to analyze modern societies in terms of their structure at any given moment. It analyzes and interprets modern societies in terms of the future of the human race. It contains within itself a philosophy of history.

Synthetic and historical, Marxist sociology is also determinist; it predicts as inevitable the advent of a certain economic and social mode of organization. It is also progressive as well as determinist, because it assumes the superiority of the coming social regime, as compared with the social regimes of the past or present.

3

This sociology, which claims to derive from Marx and which does retain many of his essential ideas, albeit in simplified form, contains two kinds of contradictory implications, depending on whether one is considering Soviet or non-Soviet societies.

The Marxist sociology of the nineteenth century was revolutionary. By this I mean that it predicted the revolution which would destroy the capitalist regime and hailed this revolution as a good thing. But today Soviet sociology regards the beneficial revolution as a thing of the past, not of the future. The anniversary of the 1917 revolution is celebrated: the revolution is forty-eight years old; and because the decisive break has occurred in the course of history, the same sociology which was revolutionary in the nineteenth century plays a conservative role today, at least in relation to Soviet society. A sociology born of a revolutionary intent has become an ideology which justifies the existing Soviet society.

But at the same time Soviet sociology has remained revolutionary in relation to non-Soviet societies. It implies that other types of societies have yet to undergo the revolution which Soviet society accomplished in 1917.

From this duality there derives a certain peculiarity of the sociological literature of the Soviet Union. The more Soviet sociologists know about their own society, the more they praise it; the less they know of other societies, the more they criticize them.

Indeed, until quite recently, Soviet sociologists could not leave the Soviet Union. But thanks to their theories they thought themselves familiar with non-Soviet society, they denounced its decadence, and they predicted its inevitable collapse. Meanwhile, dedicated in theory to the study of Soviet society, they were embarrassed in fact because they knew in advance what this postrevolutionary society was like, because they knew, at least in broad outline, what it had to be according to the laws of history.

What this comes down to is that in a postrevolutionary society, the formerly revolutionary ideology has become official; it has become the expression of the truth of the state. This clearly helps give Soviet sociologists that sense of their own importance which is lacking in Western societies. But equally clearly, a position as society's official interpreter entails certain restrictions in the search for truth.

American sociology reveals, in general, exactly opposite characteristics. American sociologists, in my own experience, never talk about laws of history, first of all because they are not acquainted with them, and next because they do not believe in their existence. Because they are men of intelligence, they would prefer to say that these laws have not been established with any certainty; but if they were to express their real thoughts, they would probably say that what the Soviets regard as laws of history have no scientific validity and that there is no justification for deriving such laws either from recurrent patterns through the ages or from evolutionary tendencies sufficiently obvious to permit the accurate prediction of the future from the present.

American sociology is fundamentally analytical and empirical; it proposes to examine the way of life of individuals in the societies with which we are familiar. Its energetic research is aimed at determining the thoughts and reactions of students in a classroom, professors in or outside their universities, workers in a factory, voters on election day, and so forth. American sociology prefers to explain institutions and structures in terms of the behavior of individuals and of the goals, mental states, and motives which determine the behavior of members of the various social groups.

This empirical or analytical sociology is not a state ideology, nor is it a conscious glorification of American society. As a matter of fact, it is carried on for the most part by persons belonging to what is known in the United States as the liberal school of political opinion. By and large, American liberals are more apt to vote for the Democratic than the Republican Party (although there are liberals in the Republican Party, too). On most of the political issues raised in the United States, they favor the solutions advocated by the European left. For example, they prefer a maximum of social mobility to the hereditary perpetuation of class distinctions.

American sociologists are more sympathetic than hostile to American society, but their sympathy takes the form of approval of the whole and criticism of the parts—which is as good a definition as any of what in Europe is called "reformism."

Soviet sociologists are conservative for themselves and revolutionaries for others. American sociologists are inclined to be reformists for the whole world. They are reformists more specifically for American society, because they are primarily concerned with American society. But their method and their point of view, transported to other societies, would probably lead to the assumption of analogous positions. Not that the analytical or empirical method of study necessarily excludes the revolutionary spirit: this method may in some cases encounter obstacles which cannot be surmounted by mere reforms. But, in most cases, the adoption of American sociological method and attitude will lead to the adoption of reformist positions. If you study social organizations in detail, you will find something to improve everywhere. In order to seek a revolution—that is, a total upheaval—you must assume an over-all viewpoint, take up a synthetic method, define the essence of a given society, and reject that essence.

These two schools, which I have characterized so sketchily, do not include the whole of what is practiced all over the world under the name of sociology. But these two schools, which are the most typical ones, form the opposite poles between which fluctuates what is called sociology today. All national schools include these two approaches in varying degrees—the tendency toward the empirical and analytical study of modern society and the tendency toward synthetic analysis of modern societies in the context of world history. Whether the sociology be analytic or synthetic, contemporary or historical, it contains implications regarding the society in question. It is impossible for a study of a given society, however scientific its aims, not to contain implied approval or criticism of that society.

The social role of sociology in the two schools I have described is perfectly clear. The role of Soviet sociology is conservative-revolutionary; that of American sociology is conservative-reformist. But, in both cases, these schools regard their revolution as lying in the past rather than the future.

As for the sociologies which are neither Soviet nor American, perhaps most of their uncertainty results from the difficulty they have in situating their revolution precisely. When one knows that the revolution is either ahead or behind, one has a clear historical perspective. The European sociologies, to the extent that they are being Americanized, are in the process of destroying the idea of Revolution with a capital "R." The sociologies, like the societies, of Western Europe are approaching the American model. German sociology increasingly resembles American analytical and empirical sociology. French sociology is also beginning to approach the American model, with the same psychological repercussions. I have observed the gradual renunciation of the revolutionary ideology in young sociologists as they became researchers. They remain leftists; but when one undertakes a concrete, detailed study of social institutions, one becomes critical in detail and forgets about total negation.

Whatever the diversity of these two schools and of all the intermediary ones, they nevertheless belong to a whole which may be called sociology. Indeed, these two schools converge in the study of common problems. In both schools, social differentiation within modern societies is studied under various aspects, designated by such terms as social stratification, class struggle, and social mobility.

Still more generally, both schools hold that one essential characteristic of modern society is the rapid transformation of the society itself. Modern societies are perhaps the first in history, not just to change, but also to be aware of change as the very nature of society. While in most past societies the ideal image was of an established and unchanging order, the ideal image of modern societies is of a steady economic growth involving constant upheavals in the social organization.

Now, this phenomenon is common to Soviet and Western societies. Clearly it is one of the preferential domains of all sociological inquiry which permits us to say that sociology has retained as a preferential domain the problem of *modern* societies, as was predicted by those who were the first theorists of sociology and whom we shall be studying in the chapters that follow.

In addition, however, to this problem of modern society, sociologists do concern themselves, on the one hand, with the groups I shall call traditional—groups which have existed in all societies as well as the modern ones: the family and the state—and, on the other hand, with the relatively new groups arising within modern society: political parties, labor unions, pressure groups, and so on.

One can say, then, that whatever the method used or the viewpoint taken, sociology has as its object the analysis, understanding, and interpretation of modern societies in the framework of the permanent organization of historical societies.

Besides having this central object, sociology is oriented in two other directions. On the one hand, there are countless specialized sociologies; and, on the other, there is the attempt to comprehend units larger than the nation itself.

The specialized sociologies—the sociology of science, the sociology of language, the sociology of art, the sociology of literature—are attempts to explain the evolution of human phenomena in relation to the social milieu. I deliberately use the phrase "in relation to," for it has a double significance, referring either to the social causes or influences which favored, opposed, or modified these intellectual creations, or to the social consequences of these intellectual creations. These so-called specialized sociologies are a specific approach to the study and interpretation of realities which have a social aspect, without their essence being exhausted by their social character.

Finally, there is the effort of sociological inquiry to determine and comprehend those vast and vague realities called "civilizations" or "cultures."

These realities are not constituted by an organization, in the manner of national entities. Whether we grant or deny the reality of a "Western Civilization," there has never been an organization which established and represented the unity of that civilization. Yet the latter has nonetheless had a certain existence. The cultures Spengler or Toynbee speak of are perhaps difficult to comprehend; but they are not pure mythology.

What conclusions may be drawn from this preliminary examination into the nature of sociology? Here are a few which will serve as our point of departure.

Sociology may be said to be characterized by two specific aims which account for its nature. On the one hand, sociology lays claim to objective and scientific knowledge. On the other, what it claims to know objectively and scientifically is some vaguely defined thing we call society or societies or social phenomena. For the time being, I shall use these expressions as if they were synonymous.

In the past, there have always been studies devoted to the organization of societies. But the traditional studies of the social order were not inspired to the same extent by the two aims I have just discussed—the aim of scientific objectivity and the aim of grasping the social *as such*.

Aristotle's *Politics* is a treatise on political sociology. But the central interest, the point of reference in Aristotle's *Politics,* is the political regime and not the social organization. Thus, to my way of thinking, sociology marks a moment in man's reflection on historical reality, the moment when the concept of the social, of society, becomes the center of interest, replacing the concept of politics or of the regime or of the state.

But these two concepts, the scientific and the social, immediately give rise to two antithetical interpretations which correspond quite neatly to the characteristics of the two major schools. For the specifically social may be sought either on the level of the part, the element, or on the level of the whole, the entity. Sociology, the science of the social, may just as well be the science of the microscopic relationships between two people on the street or three dozen people in a military or academic group, as the science of society as a whole.

This fluctuation between element and entity has an analogy in the opposition between the present and the past. Of course, these two antitheses are not equivalent. One may elaborate a sociology that is both analytical and historical. There is no necessary connection between the tendency toward the modern and the tendency toward the microscopic. But, in fact, there is a certain relation all the same. For a sociology which tends toward the interpretation of society as a whole is almost irresistibly led to define social types; and the definition of social types leads to the hypothesis of an evolution from one social type to another.

If one defines the aim proper to sociology as the combination and reunion of the study of the part with the study of the whole, it is easier to explain the significant role played by problems of delimitation in sociological theory, especially in general sociology. In itself, it is not particularly interesting to know how sociology differs from history or from psychology or from economics. But if sociology's specific intention is to analyze and to comprehend the social as such, both as element and as entity, then countless questions inevitably arise concerning the relations between sociology as such and the other social disciplines. This is why we shall, in what follows, ask every sociologist to give his interpretation of the relations between sociology and the other social disciplines. The problem of delimitation is not a vain theoretical diversion; it is the expression of the central problem of sociology as an original science

—sociology whose aim is to comprehend the social as such, by which is meant either the element present in all social relations or the larger and vaguer entity embracing and uniting the various sectors of collective life.

Montesquieu

I

IT IS FASHIONABLE nowadays to regard Charles de Secondat, Baron de la Brède et de Montesquieu, as a precursor of sociology, which is justified if the founder is the man (in this case, Auguste Comte) who invented the term. On the other hand, if the sociologist is to be defined by the peculiar aims which I have suggested, then Montesquieu was much more of a sociologist than Auguste Comte. The philosophical interpretation of sociology present in *The Spirit of the Laws* is much more "modern" than the same interpretation in the writings of Auguste Comte. This does not necessarily mean that Montesquieu was superior to Auguste Comte; but it does mean that I do not consider Montesquieu a precursor of sociology, but rather one of its great theorists.

To regard Montesquieu as a sociologist is to answer questions which every historian has raised: From what discipline did Montesquieu derive? To what school does he belong?

The uncertainty as to where Montesquieu "belongs" is apparent from the organization of curricula in French schools. Montesquieu may appear simultaneously on the reading lists of the literature department, the philosophy department, and in certain cases even the history department. On a higher level, historians of ideas rank Montesquieu in turn among the men of letters, among political theorists, among legal historians, among the eighteenth-century "ideologists" who probed the foundations of

French institutions and who prepared the way for the Revolution, and even among economists.[1]

It is very true that Montesquieu was at once a writer, almost a novelist, a jurist, and a political philosopher. But there is no question that *The Spirit of the Laws* holds a central place in his work. Now, it seems obvious to me that the aim of *The Spirit of the Laws* is the aim of sociology as I have defined it.

What is this aim? Montesquieu made no secret of it. His purpose was to make history intelligible. He sought to understand historical truth. But historical truth appeared to him in the form of an almost limitless diversity of morals, customs, ideas, laws, and institutions. His inquiry's point of departure was precisely this seemingly incoherent diversity. The goal of the inquiry should have been the replacement of this incoherent diversity by a conceptual order. One might say that Montesquieu, exactly like Max Weber, wanted to proceed from the meaningless fact to an intelligible order. This attitude is precisely the one peculiar to the sociologist.

But the two terms I have just used—meaningless diversity and intelligible order—raise an obvious problem. How does one go about discovering an intelligible order? What will be the nature of this intelligible order, which is to replace the radical diversity of customs and morals? On what level and by what means does one discover the intelligible order? What are the instruments of this intelligibility?

It seems to me that in Montesquieu's works there are two answers to these questions. They are not contradictory, but are rather two stages in the same undertaking.

The first amounts to a declaration that it is not chance which rules the world. Beyond the chaos of accidents, there are underlying causes which account for the apparent absurdity of things.

Here is a passage which occurs in his *Considerations on the Causes of the Grandeur and Decadence of the Romans:*

It is not fortune which rules the world. We can ask the Romans, who had a constant series of successes when they followed a certain plan, and an uninterrupted sequence of disasters when they followed another. There are general causes, whether moral or physical [*Remember these two terms, which play an important role in Montesquieu's system*] . . . which operate in every monarchy, to bring about its rise, its duration, and its fall. All accidents are subject to these causes, and if the outcome of a single battle, i.e., a particular cause, was the ruin of a State, there was a general cause which decreed that that State was destined to perish through a single battle. In short, the main impulse carries all the particular accidents along with it.

Here is another passage which expresses the same idea, from Chapter 13 of Book X of *The Spirit of the Laws:*

It was not the affair of Poltava that ruined Charles. Had he not been destroyed at that place, he would have been in another. The casualties of fortune are easily repaired; but who can be guarded against events that incessantly arise from the nature of things?

The idea revealed in these two quotations is, it seems to me, Montesquieu's first truly sociological idea, which I should express in this way: behind the seemingly accidental course of events, we must grasp the underlying causes which account for them.

It should be emphasized that a statement of this kind does not imply that *everything* that has happened was "necessitated" by the underlying causes. Sociology is not defined at the outset by the hypothesis that accidents have no effect at all on the course of history. After all, it is a matter of fact, not of dogma, whether a particular military victory or defeat has been caused by the corruption of the state or by errors of technique or tactics. Not every military victory is proof of the greatness of a state nor is every defeat proof of its corruption.

Montesquieu's second answer seems to me more fundamentally interesting and perceptive: it is not that apparent accidents may be explained by underlying causes, but that one can organize the diversity of manners, customs, and ideas into a small number of types. Between the infinite variety of customs and the absolute unity of an ideal society, we must discover an intermediate term, namely, a small number of social types.

The preface to *The Spirit of the Laws* clearly expresses this essential idea:

> I have first of all considered mankind, and the result of my thoughts has been, that amidst such an infinite diversity of laws and manners, they were not solely conducted by the caprice of fancy.

The statement implies that the diversity of laws may be explained, with the laws peculiar to each society being determined by certain causes which sometimes operate without our being aware of them. He continues:

> I have laid down the first principles, and have found that the particular cases follow naturally from them; that the histories of all nations are only consequences of them; and that every particular law is connected with another law, or depends on some other of a more general extent.

Thus, it is possible to organize the diversity of customs in two ways: on the one hand, by ascertaining the causes underlying the particular laws observed in a given case; on the other, by discovering the principles or models which form an intermediate level between meaningless diversity and a scheme which is universally valid.

We make *development* intelligible when we reveal the underlying causes which have determined the general direction of events. We make *diversity* intelligible when we organize it within the compass of a small number of types or concepts.

What is the conceptual tool Montesquieu used to replace a meaningless diversity with an intellectual order? The question is virtually the same as the one traditionally raised by Montesquieu's commentators: What is the plan of *The Spirit of the Laws?* Does *The Spirit of the Laws* present us with an intelligible order, or is it merely a collection of more or less acute remarks on certain aspects of historical reality?

The Spirit of the Laws is divided into several parts whose apparent heterogeneity has often been remarked. There seem to me to be essentially three main divisions.

First of all, there are the first thirteen books, which develop the well-known theory of the three types of government. These are concerned with what should be called political sociology: an attempt to reduce the diversity of forms of government to a few types, each of these being defined at the same time by its nature and its origin. The second part, from Book XIV through Book XVIII, is devoted to material or physical causes, that is, to the influence of soil and climate on human beings, their manners, and their institutions. The third part, from Book XX to Book XXVI, takes up one by one the influence of social causes—trade, currency, population, religion—on manners, customs, and laws. These three parts thus seem to be (*a*) a sociology of politics and then (*b*) a sociological survey of the material and moral causes which influence social organization.

Besides these three main divisions, there remain the last books of *The Spirit of the Laws* which, devoted to an investigation of Roman feudal legislation, represent historical illustrations; and there is Book XXIX, which is difficult to relate to one of the large divisions, devoted as it is to the question of how laws should be written. This book may be regarded as a pragmatic elaboration of the conclusions deduced from scientific investigation.

There is, lastly, one book which is difficult to classify in this over-all design. It is Book XIX, dealing with the general spirit of a nation—its morals and customs. It is not concerned with a particular cause or with the political aspect of institutions, but with what might be called the unifying principle of the social entity. This book is especially important, for it is a transition or link between the political sociology of the first part of *The Spirit of the Laws* and the other two parts, which examine material and spiritual causes.

This outline of the plan of *The Spirit of the Laws* enables us to cope with the fundamental problems involved in the interpretation of Montesquieu. Historians have all been struck by the differences between the first part and the two succeeding parts. And whenever historians remark on the apparent heterogeneity of the parts of the same book, they are tempted to resort to a peculiarly historical kind of explanation: they try to find out *when* the author wrote the various sections so apparently unrelated.

Montesquieu lends himself rather easily to this historical explanation. The first books of *The Spirit of the Laws*—not Book I, but Books II to VII or VIII, the books which analyze the three types of government—are, one might say, of Aristotelian inspiration. Montesquieu wrote these first books before his trip to England, at a time when he was under the dominant influence of classical political philosophy. Aristotle's *Politics* was *the* basic book in the classical tradition; and there is no doubt that Montesquieu wrote the first books with that volume beside him. There are allusions to or comments on the *Politics* on almost every page.

The books that follow—especially the famous Book XI on the English constitution and the separation of powers —were probably written later, after the trip to England, under the influence of observations made during this trip. As for the sociological books devoted to the study of material or spiritual causes, they were probably written more slowly than the first books.

It would therefore be easy (but not really satisfactory) to regard *The Spirit of the Laws* as the juxtaposition of two ways of thinking about, and two ways of examining, reality. On the one hand, Montesquieu would be a student of the classical philosophers. As such, he developed a theory of types of government which differs on a few points from the classical theory of Aristotle but which is still within the tradition of the philosophers. At the same time, Montesquieu would be a sociologist trying to discover how religion, climate, the nature of the soil, and the size of the population influence the various aspects of collective life. The author being two men—political theorist on the one hand, sociologist on the other—*The Spirit of the Laws* would be an inconsistent book rather than a book which, though it contains sections of different dates and perhaps of different inspiration, is ordered by a ruling purpose and a system of ideas.

Before resigning ourselves to an interpretation which assumes the historian to be so much wiser than the original author, capable of perceiving immediately the contradiction that supposedly eluded the genius, we must look for the internal order which Montesquieu, wrongly or rightly, discerned in his own thought.

The central problem is that of the compatibility of the theory of the *types* of government and the theory of *causes*. To begin with, then, I shall review the essentials of Montesquieu's theory of types of government, his political theory. Next, we shall see how this theory of types of government is related to the other parts of *The Spirit of the Laws*.

Montesquieu, of course, distinguished three types of government: republic, monarchy, and despotism. Each of these types is defined with reference to two ideas, which Montesquieu called the *nature* and the *principle* of government.

The principle of government is the sentiment which must animate men within a type of government for the latter to function harmoniously. As Montesquieu expressed it, virtue is the principle of the republic—which does not mean that in a republic all men are virtuous, but that republics are prosperous only to the extent that their citizens are virtuous.[2]

The nature of government is that which determines its form; specifically, the number of individuals possessing sovereign power or, if you prefer, sovereignty. The republic is that form of government in which the people as a body, or a part of the people, possess sovereign power. (The distinction between the people as a body and only a part of the people suggests the two types of republican government, democracy and aristocracy.) The monarchic form of government is one in which a single person governs by fixed and established laws. In a despotic government, a single person, without laws or regulations, directs everything by his own will and caprice.

From these definitions, which appear in the first chapter of Book II, it is clear that the nature of a government depends not only on the number of people who hold the sovereign power but also on the manner in which this sovereign power is exercised. Monarchy and despotism are both forms of government which place the sovereign power in the hands of a single person; but in the case of the monarchical form this individual governs according to fixed and established laws, while in a despotism he governs without laws or principles. Thus we have two criteria—or, in modern jargon, two variables—to determine the nature of each government: first, who holds the sovereign power; next, by what method this sovereign power is exercised.

In addition, of course, there is the third criterion, which is that of principle, as described above. The type of government is not sufficiently defined by a quasi-legal criterion (i.e., who holds the sovereign power). Each type of government is characterized by a sentiment, without which it cannot survive and prosper.

Now, according to Montesquieu, there are three basic political sentiments, each of which assures the stability of a single type of government. The republic depends on virtue, the monarchy on honor, and despotism on fear.

The virtue of the republic is not a moral virtue, but a peculiarly political virtue. It is respect for law and the individual's dedication to the welfare of the group.

Honor is the individual's respect for what he owes to his rank. According to Montesquieu, it is philosophically a false virtue.[3]

Fear needs no definition. It is a primal and, so to speak,

subpolitical emotion. But it is one that all political theorists have discussed, because many of them, since Hobbes, have regarded it as the most human, the most basic emotion—the emotion which underlies the state itself. Montesquieu, however, was not a pessimist *à la* Hobbes. He believed that a government based on fear is fundamentally corrupt and on the threshold of political ruin.

II

IT IS IMPORTANT to note the ways in which Montesquieu's classification of the forms of government departed from the classical tradition.

Montesquieu's first originality was to regard democracy and aristocracy (which in Aristotle's classification are two separate types) as two modes of a form of government called republican. For Montesquieu, the fundamental distinction is between the republic, including the two modes of aristocracy and democracy, and the monarchy. According to Montesquieu, Aristotle was unaware of the true nature of monarchy—which is understandable since monarchy, as Montesquieu conceived of it, had been achieved only in postclassical Europe.[4]

There is an underlying reason for this departure. In Montesquieu, the distinction between forms of government is also a conscious distinction between social organizations and structures. Aristotle had created a theory of forms of government to which he had apparently assigned a general validity, but he was presupposing the Greek city-state as its social basis. Monarchy, aristocracy, and democracy constituted the three modes of political organization of the Greek city-states. It was justifiable, in this context, to distinguish types of government according to the number of persons holding the sovereign power. But this kind of analysis really implied that these three forms of government were, in modern terms, the political superstructure of a certain type of society.

In classical political philosophy, no one bothered to examine the relationship between the types of political superstructure and the social foundations. No one had clearly formulated the question of to what extent a classification of political regimes can be made without considering the organization of societies. Montesquieu's decisive contribution was precisely to combine the analysis of forms of government with the study of social organizations in such a way that each regime is also seen as a certain type of society.

How did he establish this relationship between government and society? First, Montesquieu stated explicitly in Book VIII, Chapters 16, 17, and 19, that each of these three forms of government corresponds to, or is consistent with, a certain dimension of the society under consideration. Here is Montesquieu's most typical pronouncement, from Book VIII, Chapter 16: "It is natural for a republic to have only a small territory; otherwise it cannot long subsist."

A little further on he says:

> A monarchical state ought to be of moderate extent. Were it small, it would form itself into a republic; were it very large, the nobility, possessed of great estates, far from the eye of the prince, with a private court of their own, and protected, moreover, from sudden executions by the laws and manners of the country—such a nobility, I say, might throw off their allegiance, having nothing to fear from too slow and too distant a punishment.

Finally, a third passage, from Chapter 19 of Book VIII: "A large empire supposes a despotic authority in the person who governs."

If we wished to translate these formulas into strictly logical terms, we should probably have to abandon a vocabulary of causality. We would be unable to say that once the territory of a state exceeds a certain size, despotism is inevitable, but at least we could say that there is a natural correspondence between the size of the society and the type of government. In any case, it is by means of this theory of size that Montesquieu linked the classification of governments to what is now called social morphology, in Durkheim's term.

A second idea relating the classification of governments to the analysis of societies is that of *principle* in Montesquieu's sense of the word, i.e., the sentiment indispensable to the functioning of a certain form of government.

The theory of principle obviously leads to a theory of social organization. I have observed that the principle of the republic was virtue, to which Montesquieu assigned a peculiarly political rather than moral meaning. Virtue in the republic is love of the law, dedication to the group, or patriotism, to use a modern term. In the last analysis, virtue is dependent on a certain sense of equality. A republic is a form of government in which men live by and for the group. It is a form of government in which the members of the group regard themselves as citizens and therefore, ultimately, as equals.

In contrast, the principle of monarchy is honor, that is, the sense of what each man knows he owes to his rank and station. Montesquieu expounded the theory of honor in a tone which occasionally sounds polemical and ironic:

> In monarchies policy effects great things with as little virtue as possible. Thus in the nicest machines, art has reduced the number of movements, springs, and wheels.
>
> The state subsists independently of the love of our country, of the thirst of true glory, of self-denial, of the sacrifice of our dearest interests, and of all those heroic virtues which we admire in the ancients, and to us are known only by tradition. . . .
>
> A monarchical government supposes, as we have already observed, pre-eminence and ranks, as likewise a noble descent. Now, since it is the nature of honor to aspire to preferments and titles, it is properly placed in this government.
>
> Ambition is pernicious in a republic. But in a mon-

archy it has some good effects; it gives life to the government, and is attended with this advantage, that it is in no way dangerous, because it may be continually checked [Book III, Chapter 5].

Hence we see that the essence of monarchy is *not* to require virtue of its subjects. Montesquieu was not radically original in this respect. Since men first began to reflect on politics, they have always fluctuated between two extreme positions: either that a state prospers only when the people truly desire the good of the group; or that it is impossible for the people truly to desire the good of the group, and a good government is one in which the vices of mankind conspire for the common good. Montesquieu's theory of honor is a version of this second, realistic position: the good of the group is insured, if not by the vices of the citizens, then at best by their lesser virtues.

It is my personal feeling that in Montesquieu's chapters on honor there are two dominant attitudes or tendencies: a relative devaluation of honor in comparison with true political virtue, that of the ancients and of republics; but also a positive valuation of honor as the basis of social relationships and as a shield of the state against the supreme evil of despotism.

In effect—and this brings me to a third idea—the two forms of government, republican and monarchic, differ in kind in that one is based on equality and the other on inequality, but they also have one characteristic in common: they are moderate; no one rules in an arbitrary manner or outside the law. However, when we come to the third form of government, namely, despotism, we have left moderate government behind. Montesquieu combined with the classification of the three forms of government a dualistic classification of moderate and nonmoderate governments. Republic and monarchy are moderate; despotism is not.

The republic is based on an equalitarian organization of the relations between members of the group. Monarchy is based on discrimination and inequality. Despotism marks a return to equality; but whereas republican equality is an equality of virtue and of universal participation in the sovereign power, despotic equality is an equality of fear, impotence, and nonparticipation in the sovereign power. In Montesquieu's political thought, the key antithesis is between despotism, in which every man fears every other man, and the libertarian forms of government, in which no citizen fears any other.

In despotism, Montesquieu showed political evil in its absolute form. Despotism is that form of government in which a single person governs without laws or regulations, and consequently in which every man is afraid of every other man. There remains only one limit to the ruler's absolute power: religion; and even this is a precarious protection.

Is Montesquieu's synthesis of the theory of government and the theory of society altogether satisfactory? I shall outline the points which might provide a basis for argument or criticism.

(1) That despotism is a concrete political type, in the same sense as a republic or a monarchy, is open to debate. We know that Montesquieu took his model for the republic from the ancient republics, especially the Roman republic before the great conquests. His models for the monarchy were the European monarchies (English and French) of his day. His models for despotism were the empires he referred to as Asiatic. I say he referred to them as Asiatic, because he tended to regard the Persian, Chinese, Indian, and Japanese empires as birds of a feather, so to speak. Montesquieu's knowledge of Asia was fragmentary; nevertheless, he did have access to material which might have afforded him a more complex understanding of Asiatic despotism. In a sense, we can say that Montesquieu was responsible for an attitude toward the history of Asia which has still not completely disappeared—an attitude typical of European thought, whereby Asiatic governments are viewed as being in essence despotisms which have abolished all political structure, all law and order, all moderation. Asiatic despotism, as seen by Montesquieu, is a desert of

servitude. The absolute sovereign rules alone, without laws or regulations. In certain cases, he delegates his powers to a grand vizier; but whatever the intricacies of the relations between the despot and his entourage, there are no social classes to provide an equilibrium, no ranks, no orders; neither the equivalent of the ancient virtue, nor the equivalent of the European honor; fear rules millions of human beings across these vast reaches where the state can endure only if a single man is omnipotent.

This evocation of absolute political evil, incidentally, is not without a polemical intent with respect to the European monarchies of Montesquieu's day. Let us not forget the famous remark, "All monarchies will lose their being in despotism even as rivers in the sea." Despotism is a possible outcome of monarchies, once the latter lose respect for order, for nobility, for those intermediary bodies without which social structure collapses and the absolute and arbitrary power of a single individual triumphs over all moderation.

(2) Insofar as it posits a relation between the dimensions of the territory and the form of government, Montesquieu's theory of government risks leading to a sort of fatalism.

In Montesquieu's thought, there is a fluctuation between two extremes. It would be only too easy to find a number of passages which seem to imply a sort of hierarchy: the republic being the best form of government, next monarchy, finally despotism. Were this the case, what we would have would be less a sociological analysis of the different types of government than a hierarchy of political regimes. But from another point of view, and in other passages, each form of government seems inevitably determined by a certain dimension of the social body; here we are dealing, not with a hierarchy of values, but with a ruthless determinism.

(3) The third criticism, which touches the heart of the matter, concerns the relation between political regimes and social types.

One can, of course, consider this relationship in various ways. One may regard a political regime as sufficiently defined by a single criterion. For example, the number of those holding sovereignty has sometimes been regarded as a basis for the classification of political regimes, since it is independent of history. Whatever the size or the form of the society, the political regime would be classified as a democracy, aristocracy, or monarchy, according to the number of persons holding sovereignty.

Such was the implicit assumption of classical political philosophy, insofar as the latter constructed a theory of government apart from social organization, presupposing, as it were, the extrahistorical validity of political types.

The other extreme position consists in intimately relating political regime and social type, as Montesquieu did more or less explicitly. In this case, one arrives at what Max Weber would have called three ideal types: the ancient city-state, of small dimensions, governed as republic, democracy, or aristocracy; the ideal type of European monarchy, legal and moderate, whose essence is the differentiation of ranks; and finally, the ideal type of Asiatic despotism, a state of vast dimensions under the absolute power of a single man, where religion is the sole restraint on the sovereigns' whims and equality is restored—but an equality of universal impotence.

Montesquieu's most important and valuable idea, then, is the connection established between the form of government on the one hand and the style of interpersonal relations on the other. Social life depends on the way in which power is exercised by the government, and *vice versa*. Such an idea lends itself admirably to a sociology of government. At the same time, it obviously raises a crucial question: To what extent are political regimes conceptually separable from the historical realities in which they are embodied? Montesquieu clearly stated this problem. He found no final solution; but then, it is by no means certain that anyone has to date.

Let us turn now to another aspect of Montesquieu's political sociology—the most familiar aspect, which, moreover, is closely related to the ideas I have just summarized. As a matter of fact, the distinction between moderate and nonmoderate government, which is so central to Montesquieu's thought, enables us to incorporate Book XI's reflections on England into the theory of types of government found in the early books.

The essential passage in this connection is Chapter 6 of Book XI, in which Montesquieu examined the English constitution. This chapter has become so famous that a number of English constitutionalists have interpreted their own institutions in terms of what Montesquieu said about them, and the English reading public considered that they understood themselves after finishing *The Spirit of the Laws*.[5] Needless to say, I shall not undertake here a detailed study either of the English constitution in the eighteenth century or of Montesquieu's conception of it. My main concern is to show how Montesquieu's essential ideas about England fit into his general conception of politics.

In England, Montesquieu discovered, as he said, a state whose peculiar aim was political liberty. (Indeed, in Chapter 5 of Book XI, Montesquieu suggested a very tempting idea, namely, that each state has a mission, so to speak, a true vocation.) This aim was achieved through *representation*. Now, in Montesquieu's theory of the republic, the idea of representation does not have a position of prime importance. The republics he had in mind were the ancient republics, in which there existed an assembly of the people. The people themselves did not decide matters in detail; but they did decide important matters. When Montesquieu recalled the legislative power of the ancient city-states, he pictured an assembly *of* the people, and not an assembly elected *by* the people. It was in England that Montesquieu discovered the institution of representation.

The dominant characteristic of this government, whose goal was liberty and in which the people were represented in assemblies, is what has been called the "separation of powers," a doctrine which, as you know, has remained quite contemporary and about which there continues to be endless speculation. Montesquieu found that in England a monarch was possessed of the executive power. Because the latter required swift decision and action, it had to be in the hands of one man. The legislative power was embodied in two assemblies: the House of Lords, which represented the nobility, and the House of Commons, which represented the people. These two powers, executive and legislative, were possessed by distinct persons or bodies or organs. But Montesquieu put just as much emphasis on the co-operation between these organs as he did on their separation. In effect, he showed what each of these powers may and must do in relation to the other.

There is, of course, a third power, the judiciary. But in this sixth chapter Montesquieu said at least twice that this power is virtually null and invisible, which seems to indicate that the judiciary power ought to be essentially an interpreter of the law, that it should have as little initiative and personality as possible. It is not the power of persons, but the power of laws.

The legislative power co-operates with the executive power; it must be able to ascertain to what extent the laws have been correctly applied by the latter. As for the executive power, Montesquieu said it should not be expected to enter into the discussion of affairs but ought to have a co-operative relationship with the legislative power through what he calls the executive's right of hindrance (*faculté d'empêcher*). Montesquieu specified further that the budget must be voted on annually: "If the legislative power were to settle the subsidies, not from year to year, but forever, it would risk losing its liberty." Which clearly indicates that the annual vote on the budget is, as it were, a condition of liberty.

Chapter 6, Book XI, has been compared with passages from Locke on the same subject; and it would be tempting, if time permitted, to explain certain peculiarities or oddities in Montesquieu's account by referring to Locke.[6] For ex-

ample, at the beginning of Chapter 6 there are two definitions of the executive power: it is first defined as the power
determining matters having to do with international law,
which seems to limit it to foreign policy; but a little further
on it is defined as the power carrying out public resolutions,
which gives it another dimension. This peculiarity arises
from the fact that in one instance Montesquieu followed a
passage from Locke. But there is a fundamental difference
of purpose between Locke and Montesquieu. Whether or
not he wished to justify the revolution of 1688, Locke's
aim was to limit the royal power, to show that if the monarch exceeds certain limits, or fails in certain obligations,
the people, as the true source of sovereignty, have the
right to take action. On the other hand, Montesquieu's
essential concept was not the separation of powers, in the
juridical sense of the phrase, but what might be called the
balance of social powers, the condition of political liberty.

Throughout his analysis of the English constitution, Montesquieu presupposed a nobility; he assumed two houses,
one representing the people and the other the aristocracy.
He insisted that the nobles be judged only by their peers;
for were they to be judged otherwise, their judges might
treat them unjustly out of envy. In other words, Montesquieu, in his analysis of the English constitution, was trying
to prove that social discrimination—distinctions of class and
rank consistent with the essence of monarchy as he has
defined it—are indispensable to the balance of power.

I would say without hesitation that for Montesquieu, a
state is free when one power checks another. What justifies
this interpretation most strikingly is that in Book XI, after
he had completed his analysis of the English constitution,
Montesquieu turned back to Rome and analyzed the whole
of Roman history in terms of the relationship between the
plebs and the patriciate. What in fact interested him is the
rivalry, the competition, between the social classes which
is a condition of a moderate government precisely because
the different classes are able to balance each other.

As for the constitution itself, it is of course true that

Montesquieu showed in detail how each of the powers has such and such a right and how the various powers must co-operate. But this constitutional formalization is nothing more than the expression of a free state or, as I see it, of a free society in which no power can be abused because it is checked by other powers.

I should like to quote a passage from the *Considerations on the Causes of the Grandeur and Decadence of the Romans* which, in my opinion, perfectly summarizes Montesquieu's central theme:

> Whenever we shall find everyone at peace, in a State which calls itself a Republic, we can be sure that there is no liberty there. What we call union in a political body is a very ambiguous thing. True union is a union of harmony which causes all parties, however hostile they may seem to be, to contribute to the general good of the society, as dissonances in music contribute to the harmony of the whole. There may be union in a State where we seem to see nothing but dissension, that is, a harmony which produces happiness, which alone is true peace. It is as with the members of this universe, forever bound together by the action of some and the reaction of others.

Montesquieu's idea of social "consensus," as we would say since Auguste Comte, was that of a balance of power, or of peace established by action and reaction between social groups.[7]

If this analysis is correct, the theory of the English constitution is central to Montesquieu's political sociology, not necessarily because he saw the British constitution as a model for all countries, but because in the constitutional machinery of a monarchy of his own day he found the basis of a moderate and free state, as a result of the balance of social classes and of political powers. A model of liberty, this constitution is still aristocratic, and of this fact various explanations have been offered.

The first explanation, which has long been the view of jurists, is a theory of the juridically conceived separation of powers within the republican government. Certain well-defined rights are granted to the President of the republic and the prime minister, on the one hand, and to the assembly or assemblies, on the other. A balance is obtained by a precise ordering of the relations between the various organs.[8]

The second interpretation insists, as I have just done, on the balance of social power, but it also stresses the aristocratic temper of Montesquieu's mind and his desire to justify these intermediary bodies of the eighteenth century at a time when the latter were about to disappear. From this point of view, Montesquieu was a representative of the aristocracy inveighing against the power of the monarch in the name of his class, which is a class condemned. Caught in the trap of history, he took issue with the king on behalf of the nobility, but his polemics served only to advance the cause of the people, and not that of the aristocracy.[9]

I believe in a third explanation which continues and transcends the second in the manner of Hegel's *aufheben;* that is, it transcends while still retaining the kernel of truth.

The fact is that Montesquieu conceived of the balance of social power, the condition of liberty, only on the model of an aristocratic society. He believed that good governments were moderate and that governments could not be moderate unless power were checked by power or unless no citizen stood in fear of any other citizen. The nobles could have a sense of security only if their rights were guaranteed by the political organization itself. And only if the nobles felt secure could monarch and people feel equally secure. The idea of social balance expounded by Montesquieu is quite definitely linked to an aristocratic society.

It remains to be seen whether Montesquieu's conception of the conditions of liberty and of moderation apply outside the aristocratic model he had in mind. Montesquieu would probably have said that we can, of course, imagine a social metamorphosis which would tend to do away with the differentiation of ranks and orders. But can we imagine a society without ranks or orders, a state without a plurality of powers, which would at the same time be moderate and whose citizens would be free? History has shown that a democratic regime, in which the sovereignty belongs to all, is not necessarily a moderate and free government. It seems to me that Montesquieu was perfectly right in maintaining the fundamental distinction between the power of the people and the liberty of the citizens. It may be that when the people are sovereign, the security of the citizens and moderation in the exercise of power disappear.

In other words, beyond his particular aristocratic conception of the balance of social powers and the co-operation of political powers,[10] Montesquieu enunciated this general principle: the condition for respect of the laws and security of the citizens is that no power be unlimited. That is what I regard as the essential theme of what might be called Montesquieu's political sociology.

III

WE NOW APPROACH three critical problems of sociology. First, using Montesquieu's political sociology as the point of departure, how does the political sociology of *The Spirit of the Laws* relate to the sociology of the social entity? How do we effect the transition from a study of the type of government to a grasp of the society as a whole?

Second, from the standpoint of political sociology, what is the relation between fact and value, between the understanding of institutions and the determination of the good or desirable regime? We have seen that Montesquieu condemned despotism as evil, contrary to human nature. Yet he explained that despotism is more or less inevitable as the result of the size of the collectivity and of certain external conditions. Whence a problem for sociologists: how can one explain certain institutions as predetermined, independent of human will, and at the same time apply value judgments to these institutions? Does this not lead to a kind of contradiction which consists in recognizing as inevitable under certain conditions a regime which one condemns as inhuman?

Nor do we need force the passages unduly to find the contradiction. In certain places Montesquieu writes: when the dimensions of the territory are too vast, the power must be concentrated in the hands of a single person—a statement which found favor with Catherine II of Russia, for immediately apparent reasons. Elsewhere we find an even more categorical proposition: that despotism is contrary to human nature. Can a sociologist maintain that a form of government which under certain conditions is inevitable is contrary to human nature? This is the second problem.

Finally, the third problem will be the relation between rational universalism and the particularities of history. To illustrate this, let us return to the example of despotism. Montesquieu said that despotism is contrary to human nature. This raises these questions: What is human nature, that nature common to all men in all ages and all climes? How far do the peculiarly "human" characteristics extend and how are we to reconcile this appeal to human nature with the recognition of the infinite variety of morals, manners, and institutions?

Our first problem may itself be dissolved into three questions: (1) What causes that are external to the political regime did Montesquieu discuss? (2) What is the nature of the relation he established between the causes and the phenomena to be explained? (3) Does *The Spirit of the Laws* contain a synthetic interpretation of society considered as a whole, or is it merely an enumeration of causes and a juxtaposition of various relations between certain causes and effects without any of these causes or determinants being decisive?

On the first point, the enumeration of causes, I can be brief, since I have already discussed these. Moreover, at first glance, Montesquieu's treatment has nothing systematic about it. The causes he listed are as follows:

First of all, Montesquieu studied what we call the influence of the geographical milieu, the latter being subdivided, as it were, into two parts: the climate and the soil. He first studied the climate and its influence on the way of life and social institutions of the people. Next, he considered the nature of the land and tried to discover in a similar fashion how the people have cultivated the soil and distributed property as a result of its nature.

After the influence of geographical milieu comes, in Book XIX, the consideration of the general spirit of a nation, which will concern us later, for the phrase itself is ambiguous.

Then Montesquieu turned from the physical to the social causes, among which figure prominently trade, the historical revolutions in trade, and currency. One might say that he dealt essentially with the economic aspect of collective life, were it not for the fact that he virtually ignored one element which for us is essential in the analysis of the economy, namely, the means of production, to use the Marxist term, or the technical tools and instruments men have at their disposal.

The essence of the economy for Montesquieu is either the distribution of property, especially land; or trade, exchanges, and communications between collectivities; or, finally, currency, which he saw as an essential aspect of the relations between men within collectivities or between collectivities. Montesquieu saw the economy as divided between agriculture and trade; not that he ignored what he called "the arts," the beginnings of what we call industry, but in his eyes the cities in which the economic interest prevails are the mercantile or commercial cities: Athens, Venice, Genoa. In other words, the essential antithesis is between collectivities whose chief preoccupation is military activity, glory, and the collectivities whose chief preoccupation is trade.

I want to dwell for a moment on this idea, which is a traditional one in premodern political philosophy. The originality of modern societies, which is linked to industry, was not apparent to classical political philosophy, and in this respect Montesquieu belongs to the classical tradition. We can even say that he is anterior to the Encyclopedists, that he was far from grasping the implications of the technological discoveries for the transformation of society as a whole.

After trade and currency comes the analysis of population. In history, the problem arises in two forms. In certain cases, there is a struggle against depopulation—the more common condition, according to Montesquieu; in his opinion, the danger threatening most societies is lack of manpower. But he also recognized the converse problem, the struggle against overpopulation, against the growth of population beyond available resources.

Finally, after population comes religion, which Montesquieu considered one of the most powerful influences on the organization of collective life.

As we see, this enumeration appears to have no systematic character. The author considered a certain number of causes in turn. His dominant distinction seems to be that of physical causes and moral causes. Climate and the nature of the land belong to the physical causes. The general spirit of a nation and religion belong to the moral causes. As for trade and population, it would have been easy to make a single category of these, grouping the characteristics of collective life which influence other aspects of this same collective life. But Montesquieu did not construct a systematic scheme of this kind.

However, we need only change the order to have a reasonably satisfactory list. Starting from the geographical milieu, with the two concepts of climate and the nature of the land, we would proceed to population, because it would be more logical to go from the physical milieu, which limits the size of the society, to the number of inhabitants. From here we would consider the strictly social causes, of which Montesquieu recognized at least two major ones: the body of beliefs which he called religion and the organization of labor and of trade. We would conclude with what is the true culmination of Montesquieu's sociology, namely, the general spirit of a nation.

As for the determinates, i.e., those phenomena which Montesquieu sought to explain by means of the causes which he studied in turn, it seems to me that Montesquieu used three concepts—laws, manners, and customs—which he carefully defined in Chapter 16 of Book XIX:

> Manners and customs are those usages which laws have not established, either because they were not able or were not willing to establish them.
> There is this difference between laws and manners, that the laws are most adapted to regulate the actions of the subject, and manners to regulate the actions of man. There is this difference between manners and customs, that the former principally relate to internal behavior, the latter to external behavior.

The first distinction, between laws and manners, corresponds to the sociologist's distinction between what is decreed by the state and what is imposed by society. In one case, there are explicitly worded mandates sanctioned by the state itself. In the other—the case of manners—there are both positive and negative mandates, orders or prohibitions, which are imposed on the members of a group without any law enforcing their respect and without any officially provided punishments in the event of violation.

The distinction between manners and customs corresponds to the distinction between internalized imperatives and purely external ways of behaving, prescribed by the group.

As for the laws themselves, Montesquieu distinguished the various types, essentially the three principal ones: civil laws, which have to do with the organization of family life; criminal laws, in which, like all his contemporaries, he was passionately interested;[11] and those laws directly related to the political regime.

Let us turn now to our second question, namely, the nature of the relation established between determinants and determinates, or between the causes and the institutions to be explained. For illustration, I shall take the books dealing with geographical milieu—that is, with climate and land—celebrated books in which I feel the personal peculiarities of Montesquieu's theories are most readily perceived.

As we have said, Montesquieu was primarily concerned, in treating the geographical milieu, with climate and land, but his development of these notions is quite summary. As regards climate, the distinction is virtually reduced to the contrast between hot and cold, temperate and extreme. Of course, modern geographers utilize concepts far more precise than these, with a great many distinctions between the different types of climate. As regards land, Montesquieu chiefly considered its fertility or sterility, more briefly its relief, the plains and mountains, and the distribution of relief over a given continent. On all these points, Montesquieu was not particularly original; he borrowed many of his geographical ideas from the English doctor, Arbuthnot.[12] M. Dedieu, who is one of the most qualified of Montesquieu's commentators, has read all the English literature on the subject and has proven its influence on Montesquieu. But these questions of historical erudition do not concern me. What does interest me is the logical nature of the causal relations posited by Montesquieu.

In many cases, Montesquieu directly attributed the temperaments of men, their sensibility, their way of life, to climate. Here is the typical formula, from Chapter 2 of Book XIV:

> In cold countries men have very little sensibility for the pleasures of life; in temperate countries, they have more; in warm countries, their sensibility is exquisite.

Sociology would be simple if statements of this order were true. And consider this remark of Montesquieu's:

> As climates are distinguished by degrees of latitude, we might distinguish them also in some measure by those of sensibility. . . . I have been at the opera in England and in Italy, where I have seen the same

38

pieces and the same performers; and yet the same music produces such different effects on the two nations: one is so cold and phlegmatic, and the other so lively and enraptured, that it seems almost inconceivable.

Logically, then, the proposition is of the following type: a certain physical environment is directly responsible for certain physiological, nervous, and psychological traits in the inhabitants.

But there are more complex explanations as well. I shall take a famous example—that of slavery—from Book XV, in which Montesquieu dealt with the relation between slavery and climate, and which even bears the title, "In What Manner the Laws of Civil Slavery Relate to the Nature of the Climate":

> There are countries where the excess of heat enervates the body, and renders men so slothful and dispirited that nothing but the fear of chastisement can oblige them to perform any laborious duty: slavery is there less offensive to reason; and the master being as lazy with respect to his sovereign as his slave is with regard to him, this adds a political to a civil slavery.

A passage of this kind is interesting because it reveals different facets of Montesquieu's mind. At first glance, it offers a simple, almost oversimplified explanation of the relation between climate and slavery. But in this same passage we find the statement, "slavery there is less offensive to reason," which implies that slavery by its very nature is offensive to reason and contains an implicit reference to a universal conception of human nature. Reason excludes certain institutions, or at any rate considers them evil. Here we find, juxtaposed, the two aspects of Montesquieu's interpretation: the determinist view of institutions as facts, on the one hand, and on the other a judgment on those institutions in the name of universally valid values.

The compatibility of these two modes of thought is achieved, in this case, by the phrase "is less offensive to reason." Although he admits that slavery by its spirit is contrary to the essence of human nature, Montesquieu found a justification for the reality of slavery in the influence of the climate. But a statement of this sort is logically admissible only to the extent that the climate exerts an influence on, or is favorable to, an institution without making it inevitable. For were this an inexorable relation of cause and effect, then we should have a pure contradiction between a moral condemnation and a scientifically demonstrated determinism.

I find confirmation of the nature of this causal relation in Chapter 8 of the same Book XV. Montesquieu concluded with these lines, which again are typical:

> I know not whether this article be dictated by my understanding or by my heart. Possibly there is not that climate upon earth where the most laborious services might not with proper encouragement be performed by free men. Bad laws having made lazy men, they have been reduced to slavery because of their laziness.

In appearance, a passage of this kind seems to contradict the passage quoted above. The first passage seems to attribute slavery to the climate. The second attributes it to bad laws and offers the proposition, "nowhere is the climate such that slavery was inevitable."

Montesquieu was in a dilemma, as are all sociologists when they encounter phenomena of this sort. If they carry the deterministic interpretation to its logical conclusion and find that the institution which is repugnant to them was inevitable, they must accept the consequences. We are still safe as long as we are dealing with institutions of past ages, for the past is unalterable; we need not wonder about what might have been. But if we apply these ideas to present societies (and if we apply them to past societies, why not to present ones?), we reach an impasse: how can the sociologist suggest reform if the most inhuman institutions are inevitable?

One way out of this dilemma is to locate explanations

of institutions in terms of their geographical milieu within a category which a modern sociologist would call, not a relation of causal necessity, but a relation of influence. A certain cause makes one institution not inevitable, but more probable than another. And, indeed, from his analysis of these influences, Montesquieu seems to have concluded that the legislator's task is often to counteract the direct influence of the natural phenomena. Chapter 5 of Book XIV is entitled, "That those are bad Legislators who favor the Vices of the Climate, and good Legislators who oppose those Vices." The legislator's task is therefore to introduce into the fabric of determinism human laws which oppose the direct, immediate influence of the natural phenomena.[13]

Montesquieu was therefore less an advocate of strict climatic determinism than has been maintained. The fact is that, like many of his contemporaries, he recognized in an oversimplified manner that men's temperaments and sensibilities were directly influenced by the climate. He tried, moreover, to establish a relation—of probability, let us call it—between physical realities and certain social institutions. But since he recognized the multiplicity of causes and the legislator's possible influence, the tenor of his thought is that milieu does not determine social organization but exerts an influence on it and helps orient it in a certain direction.

Apart from these generalities, to digress for a moment, Montesquieu had a number of curious and amusing observations. In yet another illustration, Chapter 13 of Book XIV, Montesquieu, still preoccupied with England, attempted to attribute the peculiarities of English life to the climate of the British Isles. And, as we see, he did not have an easy time of it:

> In a nation so distempered by the climate as to have a disrelish of everything, nay, even of life, it is plain that the government most suitable to the inhabitants is that in which they cannot lay their uneasiness to any single person's charge, and in which, being under the direction rather of the laws than of the prince, it is

impossible for them to change the government without subverting the laws themselves.

This difficult sentence seems to mean that the climate of England drives the inhabitants to such despair that it has been necessary to abolish government by a single person so that the natural bitterness of the inhabitants of the British Isles can be vented on the intangible laws rather than on the person of a single man.

The analysis of the English climate continues in this vein for several paragraphs. Here is another sample:

And if this nation has likewise derived from the climate a certain impatience of temper, which renders them incapable of bearing the same train of things for any long continuance, it is obvious that the government above mentioned is the fittest for them.

Thus the impatience of the British people is in subtle harmony with a government whose subjects, unable to ascribe their grievances to a single possessor of power, are in a sense paralyzed in the expression of their impatience.

In the books on climate, Montesquieu abounded in formulas of this sort, which are ingenious, though hardly convincing.

Now a few words about another of the determinants, namely, the number of inhabitants, or size of the population. We must turn to Book XXIII, in which Montesquieu examined in turn the problems raised, first by societies threatened with depopulation, next by societies which, on the contrary, are threatened with overpopulation. In this book, I call attention to only one curious chapter which reveals some of Montesquieu's key ideas. It is Chapter 15, in which he stated the problem of the relation of population to the arts. The word *art* is used in the sense of the activity peculiar to artisans. Thus, we are dealing with activities such as the production and modification of objects (artifacts), and not with the direct cultivation of the soil. In this chapter, Montesquieu raised what is for us the fundamental problem of population size, which is of course influenced by the means of production and the organization of labor.

In a general sense, population size is dependent on the potentialities of agricultural production. There may be as many people in a given culture as the farmers can feed. However, if the land is cultivated more efficiently, the farmers are able to feed not only themselves, but others as well. But the farmers must be willing to produce more than is necessary for their own subsistence. This raises the problem of motivating the farmers to maximum production and of stimulating the exchange of goods produced in the cities by the arts for goods produced by the farmers.

Montesquieu arrived at the idea that in order to motivate the peasants to produce, they must be given a desire for luxury. Once again, this is sound theory. One can set in motion the process of expansion in underdeveloped cultures only by creating needs in farmers whose living conditions are traditional. They must desire to possess more than they are accustomed to having. Now, Montesquieu tells us, only the artisans can provide this luxury. But he continues:

> The machines designed to abridge art are not always useful. If a piece of workmanship is of a moderate price, such as is equally agreeable to the maker and the buyer, those machines which would render the manufacture more simple, or, in other words, diminish the number of workmen, would be pernicious. And if water-mills were not everywhere established, I should not have believed them so useful as is pretended, because they have deprived an infinite multitude of their employment, a vast number of persons of the use of water, and great part of the land of its fertility.

I find this passage extremely interesting. The machines "designed to abridge art," in a modern terminology inferior to Montesquieu's, are the machines which reduce the working time required to produce manufactured objects. What Montesquieu was worried about is what we call technological unemployment. If, with the help of a machine, the same object can be produced in less working time, it will be necessary to dismiss a certain number of workers. Montesquieu was worried about this, just as men of every generation have worried about it for two hundred years.

What is omitted from this line of argument? An idea that has come to be the basis of all modern economics, namely, the idea of productivity. If the same object is produced in less working time, the workers thus freed can be employed for other jobs, and thus the volume of products available to the whole group is increased. Montesquieu clearly lacked an element of economic doctrine which was not unknown to his age (the Encyclopedists understood it): he did not understand the economic significance of scientific and technological progress. This lacuna is rather strange, for Montesquieu was very much interested in the arts and sciences. He wrote numerous essays on the sciences and on technological discoveries. But he did not grasp the mechanism whereby curtailment of the working time required for a given act of production makes it possible to employ more workers and to increase the over-all volume of production.[14]

This brings me to the third question, namely, to what degree did Montesquieu transcend analytical sociology, multiplicity of causes, partial determinants, and partial determinates? How did he succeed in reconstructing the whole?

Insofar as Montesquieu had a synthetic conception of society, I think it is found in Book XIX. I refer to Chapter 4 of this book, where we find the definition of the general spirit of a nation:

> Mankind is influenced by various causes: by climate, by religion, by laws, by maxims of government, by precedents, morals, and customs; whence is formed a general spirit of nations.
> In proportion as, in every country, any one of these causes acts with more force, the others in the same

degree are weakened. Nature and climate rule almost alone over the savages; customs govern the Chinese; the laws tyrannize in Japan; morals had formerly all their influence in Sparta; maxims of government, and the ancient simplicity of manners once prevailed at Rome.

The general spirit, then, is not a partial cause comparable to the others, but a product of that totality of physical, social, and moral causes. But it is a product which enables us to understand what constitutes the originality and unity of a given collectivity. There is a general spirit of France, a general spirit of England. We can proceed from the diversity of causes to the unity of the general spirit without the latter's excluding the former. The general spirit is not a ruling, all-powerful cause which would do away with all the others. It is, rather, that quality which a given collectivity acquires over a period of time as a result of the variety of influences exerted on it.

Montesquieu added another proposition, namely, that in the course of history a particular cause may gradually become predominant. Specifically, he formulated a theory which is still classic today: that in archaic societies the predominance of physical factors is more compelling than in more complex or, as we say and as Montesquieu would have said, more civilized societies. Montesquieu would probably have agreed that in the case of ancient nations like France or England, the influence of such physical factors as climate or soil is weak in comparison with the influence of moral causes.

In short, the theory of the general spirit leads from the sociology of politics to the sociology of the social entity. As a matter of fact, the general spirit of a nation is closely related to Montesquieu's "principle" of government. The principle is the sentiment which sustains a political regime, and this sentiment is in turn closely related to a nation's way of life as expressed by its institutions. I am inclined to think that what Montesquieu called the general spirit of a nation is what the American anthropologists call the "culture" of a nation, i.e., a certain style of life and human relations which is less a cause than a result—the result of the totality of physical and moral influences which have shaped the collectivity down through the ages.

IV

BEFORE PROCEEDING to the second division of Montesquieu's thought, I should like to raise a couple of points relative to the matters discussed earlier.

First of all, it must be borne in mind that my summaries of Montesquieu's views on geography and population were deliberately oversimplified. What is to be found in his book is much fuller, subtler, and more precise than the summary I provided. For example, I commented on Montesquieu's seeming lack of understanding of the basic phenomenon of productivity, since he emphasized the unemployment which may result from the introduction of a new technological process. It would be unfair and ridiculous to reduce Montesquieu's economic analyses to this single mistake. Actually, Montesquieu presented a picture of the factors which influence the growth of economies that is quite detailed and generally accurate.

As an economist, Montesquieu was not particularly systematic. He belonged to neither the mercantilist nor the physiocrat school. But he may be regarded, as he has been recently, as a sociologist who anticipated the modern analysis of economic development, precisely because he took into consideration the multiple factors involved. He distinguished between systems of land ownership and traced the consequences of different systems of ownership on the number of workers and the productivity of the culture; he related the system of ownership and agricultural labor to population size. Next, he related population size to the diversity of social classes. He outlined a theory which might be called the theory of luxury: the rich are necessary as a market for useless objects—objects which answer no compelling need of existence. He related domestic trade between the various social classes to trade outside the collectivity. He took up the subject of currency and traces its role in intra- and intercollectivity transactions. Finally, he inquired to what extent a given political regime is favorable or unfavorable to economic prosperity.

We see, then, that Montesquieu's is a more general, less schematic analysis than that of the economists, in the strict sense of the word. Montesquieu's ambition, indeed, was to arrive at a general sociology which would encompass economic theory. In this type of analysis, there is a constant interplay of the various elements. The mode of land ownership influences the nature of agricultural labor, and the latter in turn influences the relations between social classes. The structure of the social classes has its effect on trade, both foreign and domestic. In other words, logically, the central idea is the endless interaction of the different sectors of the social entity.

My second observation concerns the concept of a general spirit which is, as you know, the only synthetic idea to be found in Montesquieu's general sociology—except, of course, for the idea of the dominant influence of the political regime outlined in the first pages of *The Spirit of the Laws*. The two synthetic ideas are related, because a regime endures only so long as the necessary sentiment exists in the people. The general spirit of a nation is that which best contributes to sustaining this sentiment or principle which is indispensable to the continuation of the regime.

The general spirit of a nation is not comparable to the creative will of an individual or a group. It does not resemble the existential choice of a Kant or a Sartre, a single decision which is the source of the variety of actions or episodes of individual or collective life. The general spirit of a nation is the way of living, behaving, thinking, and feeling of a particular collectivity, as geography and history have produced it.

The concept of a general spirit permits the regrouping

of all partial determinants without representing itself as an ultimate explanation which would encompass all the others. Thus, in Chapter 5 of Book XVIII, Montesquieu writes:

> The inhabitants of islands have a higher relish for liberty than those of the continent. Islands are commonly of small extent; one part of the people cannot be so easily employed to oppress the other; the sea separates them from great empires; tyranny cannot so well support itself within a small compass; conquerors are stopped by the sea; and the islanders, being outside the reach of their arms, more easily preserve their own laws.

Some of these statements are debatable, but we are here concerned exclusively with Montesquieu's method. And in this chapter we see how a certain geographical situation favors one type of political institution without determining it.

A second illustrative text is much too long to quote. It is Chapter 27 of Book XIX, entitled, "How the Laws contribute to form the Manners, Customs, and Character of a Nation." The subject, of course, is England. This chapter completes Chapter 6 of Book XI, devoted to the analysis of the English constitution. I suggest reading these two chapters together, for they reveal not only how the theory of political sentiment or principle connects with the theory of the general spirit of a nation, but how the multiple, partial determinants may be regrouped into this over-all interpretation without negating the plurality of the partial explanations.

I should like to turn now to the last part of this rapid survey of Montesquieu and take up another problem, also a classic one in sociological literature, namely, the relation between facts and values and between particulars and universals. These two polarities are not identical, but they do intersect, as we shall see, and they represent two problems fundamental to any historical sociology.

The first problem could be stated this way: Is the sociologist doomed to observe the diversity of institutions without making value judgments on their merits? Must he analyze slavery, as well as liberal institutions, without being able to set up a means of discriminating among the moral or human merits of these institutions?

The second problem is: To the degree that he perceives a diversity of institutions, is the sociologist obliged to analyze this diversity without incorporating it into a system, or can he find common elements beyond the diversity?

These two polarities, without being identical, may yet coincide if the criteria which determine our value judgments are at the same time criteria of *universal* validity.

I think we can best approach these problems by going back to the central idea of *The Spirit of the Laws,* which up to now I have deliberately avoided. After all, the title of Montesquieu's great book is *The Spirit of the Laws,* and it is in the analysis of the idea or ideas of law that we find the answer to the problems I have just stated.

To modern minds, influenced by the philosophy of Kant and by logic as it is taught in our universities, the word *law* has two meanings. Law is, first of all, a command of the legislator, an order issued by a qualified authority, which compels us to do this or not to do that. Let us call this first meaning the law-as-command and go on to say that this positive law, the law of the legislator, differs from manners and customs in that it is explicitly formulated, while the obligations or prohibitions of custom are not elaborated or codified, nor do they generally carry the same type of official sanction.

Secondly, law can be taken to mean a causal relation between a determinant and an effect. For example, if we assert that slavery is a necessary consequence of a certain climate, we have a causal law which establishes a permanent relationship between a geographical milieu of a fixed type and a particular social institution.

Montesquieu claimed that he was not discussing laws, but the spirit of laws. Here is the passage, from the end of Chapter 3 of Book I:

They should be in relation to the climate of each country, to the quality of its soil, to its situation and extent, to the principal occupation of the natives, whether husbandmen, huntsmen, or shepherds: they should have relation to the degree of liberty which the constitution will bear; to the religion of the inhabitants, to their inclinations, riches, numbers, commerce, manners, and customs. In fine, they have relations to each other, as also to their origin, to the intent of the legislator, and to the order of things on which they are established; in all of which different lights they ought to be considered.

This is what I have undertaken to perform in the following work. These relations I shall examine, since all these together constitute what I call the Spirit of Laws.

Thus Montesquieu was seeking the causal laws which account for laws-as-commands. According to this passage, the spirit of the laws is precisely the totality of the relations of the laws-as-commands of various human societies with the factors capable of influencing or determining them. So apparently, in this sense, *the spirit of the laws* refers to the totality of causal relations accounting for the laws-as-commands. Since we and Montesquieu both use the word *law* in these two senses, there is a considerable likelihood of misunderstanding.

If Montesquieu's thought could merely be reduced to the above formulas, interpretation would be easy, for the laws-as-commands would be our object of study, and the causal relations would explain the laws-as-commands. If this interpretation were correct, Montesquieu would obediently conform to the portrait painted by Auguste Comte and certain modern interpreters.[15] He would be the expounder of a deterministic philosophy of law, which would observe the diversity of legislation and explain it in terms of the multiplicity of influences at work on human groups.

There are, in fact, passages in Montesquieu which tend in this direction. For example, in the Preface, Montesquieu writes:

I write not to censure anything established in any country whatsoever. Every nation will here find the reasons on which its maxims are founded; and this will be the natural inference, that to propose alterations belongs only to those who are so happy as to be born with a genius capable of penetrating the entire constitution of a state.

And, still more impressively:

Could I succeed so as to afford new reasons to every man to love his prince, his country, his laws; new reasons to render him more sensible in every nation and government of the blessings he enjoys, I should think myself the most happy of mortals.

These passages from the Preface to *The Spirit of the Laws* can be accounted for by consideration of expedience. But it is certainly true that Montesquieu could have been just as strictly conservative as his philosophy was strictly determinist. If we assume that a collectivity's institutions are inevitably determined by a body of circumstances, it is easy to shift gradually to the conclusion that the institutions are the best ones possible. It would remain to be seen whether we must add: in the best or the worst of all possible worlds.

Even the many passages in Montesquieu containing advice to legislators do not contradict the determinist philosophy I have just outlined. If you have demonstrated that legislation is a result of the spirit of a nation, it is logical to conclude that you should adapt your laws-as-commands to the spirit of the nation. There is a famous chapter on the spirit of the French nation (XIX, 5) which concludes with the advice: "Allow it to do the most frivolous things seriously, and gaily those things most serious."

Moreover, once you have reduced a regime to its nature and principle, it is easy to show what laws are suitable to the regime. For example, if a republic is based on human equality, the logical conclusion is that the laws governing education or economics must promote the sense of equality or prevent the accumulation of large fortunes.

Determinist philosophy is not incompatible with the giving of advice; but the advice must take the form of conditional or hypothetical imperatives. The legislator assumes a given set of circumstances and works out the regulations necessary to uphold the form of government suited to those circumstances or to promote the kind of prosperity a nation can possess in those circumstances. This kind of advice is on the order of what Lévy-Bruhl would have called "rational art"; it is derived from science; it unfolds the pragmatic consequences of a scientific sociology.

But there are many other passages in *The Spirit of the Laws* in which Montesquieu offered, not pragmatic advice to the legislator, but moral condemnations of certain institutions. The most celebrated passages, familiar to everyone, are those on slavery in Book XV. I also refer to Chapter 13 of Book XXV, where Montesquieu repeatedly gave free rein to his indignation against a particular form of collective organization; the chapter is entitled "A most humble Remonstrance to the Inquisitors of Spain and Portugal" and is an eloquent protest against the Inquisition.

It would be easy to explain this away by saying that Montesquieu was a man, and not merely a sociologist. As a sociologist, he justified slavery. When he was revolted by it, it was the man speaking. After all, as we have already noted, in Chapter 8 of Book XV Montesquieu writes: "I know not whether this article be dictated by my understanding or by my heart. Possibly there is not that climate upon earth where the most laborious services might not with proper encouragement be performed by free men. Bad laws having made lazy men, etc." When he condemned or defended, it is because he forgot he was writing a book on sociology.

But this explanation—i.e., dismissing moral judgments as the voice of Montesquieu the man rather than of Montesquieu the scientist—contradicts some of the most fundamental passages—those found in the first book of *The Spirit of the Laws*—where Montesquieu was constructing a theory of the various kinds of laws.

In the first chapter of Book I, Montesquieu stated explicitly that there are relations of justice and injustice anterior to positive laws: "We must therefore acknowledge relations of justice antecedent to the positive law by which they are established." And there is this other, more famous remark: "To say that there is nothing just or unjust but what is commanded or forbidden by positive laws, is the same as saying that before the describing of a circle all the radii were not equal."

In other words, if we are to take the above statement seriously, Montesquieu believed in relations of equity, in principles of justice, which are universally valid and are antecedent to positive law. What are these relations of equity? Here is the brief passage in which Montesquieu explained them:

> These relations of justice antecedent to the positive law are, for instance, if human societies existed, it would be right to conform to their laws; if there were intelligent beings that had received a benefit of another being, they ought to show their gratitude; if one intelligent being had created another intelligent being, the latter ought to continue in its original state of dependence; if one intelligent being injures another, it deserves a retaliation, and so on.

This enumeration of the relations of justice antecedent to positive law does not seem to be a systematic one. But if you read this passage attentively, you will see that in the last analysis everything can be reduced to two concepts: human equality and reciprocity. These antecedent laws of reason, these supreme laws, are based on the natural equality of men and on the obligations of reciprocity which proceed from this fundamental equality.

These antecedent laws are obviously not causal laws; they must, then, be laws-as-commands. But they are laws-as-commands that do not originate in the will of the individual legislators; rather, according to Montesquieu, they are consubstantial with nature or with human reason.

There must, then, be a third kind of law. Beyond the positive laws decreed by various societies, beyond the causal laws which establish relations among these positive laws and the influences operating on them, there must be laws-as-commands which are universally valid, whose legislator is unknown unless it be God himself, as Montesquieu seems to imply, although we cannot be sure that this is what he really means.

We have now reached the central problem in the interpretation of Montesquieu. There are those who say, in effect, that these natural laws, these laws of universally valid reason, have no place in Montesquieu's own thought. Montesquieu must have held onto them either out of caution or habit, revolutionaries always being in some respect more conservative than they think they are. What is meaningful in Montesquieu is the sociological analysis of positive law, the application of determinism to social nature. The logic of his thought consists of only three elements: the observation of the diversity of positive laws, the analysis of this diversity in terms of multiple causes, and finally the practical advice offered to the legislator as a result of the scientific exposition of the laws.

In this sense, Montesquieu would be a true positivist sociologist who tells people why they live in a certain way. The sociologist understands other men better than they understand themselves, because he discovers the causes behind the particular form that collective life assumes in different climates and ages; thus he is in a position to help the different societies live in accordance with their own nature, i.e., in accordance with their form of government, their climate, the spirit of the nation. There would be no place in this schema for the universal laws of human reason or human nature. The first chapter of Book I of *The Spirit of the Laws* would be of no importance, or rather it would be a survival in Montesquieu's doctrine of a traditional way of thinking.

Personally, I do not think this interpretation does Montesquieu justice. Moreover, I am not convinced that *anyone* has ever carried this wholly determinist philosophy of laws to its conclusion. For if we did, we could say nothing that would be universally valid in weighing the merits of a republic and a despotism. Such a radical relativism, though imaginable, seems in fact to be beyond human achievement.

What, then, is the philosophy toward which Montesquieu is moving in a more or less disorganized manner? The solution Montesquieu offered is as follows. In the first chapter of Book I, he suggested a sort of hierarchy of beings, from inorganic nature to man. Each kind of being is subject to laws. In the case of matter, these laws are purely and simply causal laws; they are inevitable laws which cannot be violated. When we come to living matter, the laws are also causal laws, but they are of a more complex nature. Finally, when we arrive at man, these laws—since, as Montesquieu said, they apply to an intelligent being—can be violated, because with intelligence comes freedom. The laws relative to human behavior are no longer in the category of inescapable causality.

Whence a formula which has always seemed paradoxical and which Léon Brunschvicg, in particular, in *Le Progrès de la Conscience dans la Philosophie Occidentale,* quotes disapprovingly:

> Indeed, the rational world must be governed just as well as the material world, for although the former also has laws which are by definition unchanging, it does not follow them faithfully as the material world follows its laws. The reason for this is that individual intelligent beings are limited by their nature, and consequently subject to error. And yet it is because of their nature that they act independently.

This passage seems to posit an inferiority of the intelligent world in relation to the material world, because the laws of the intelligent world—the rational laws governing intelligent beings—can be violated. But, as a matter of fact, there is no need to consider the possible violation of the rational laws as a proof of the inferiority of the intelligent world; quite the contrary, it may be considered the expression and the proof of human liberty.

After this metaphysic of beings comes a second chapter in *The Spirit of the Laws* entitled "Of the Laws of Nature," in which Montesquieu defined natural man—that is, his conception of man-as-man, antecedent to society, as it were. The phrase "antecedent to society" does not mean that Montesquieu believed that there were ever men who did not live in society, but that with the aid of reason we can try to conceive of what man is apart from his collective existence.

In this chapter, Montesquieu tried to refute Hobbes's theory of nature, a refutation which in my opinion provides us with a means of access to an understanding of the fundamental themes of his thought.

Actually, Montesquieu was trying to prove that man is not naturally warlike. The state of nature is not a state of universal warfare but, rather, if not a true peace, at least a state alien to the distinction between war and peace. Why was Montesquieu anxious to refute Hobbes? For the following reason: since in Hobbes's view of the state of nature, man abandoned to his desires finds himself immediately at variance with his fellow creatures, Hobbes found himself logically compelled to justify that political absolutism which alone is capable of imposing peace and affording security to a quarrelsome species. Montesquieu, on the other hand, did not find the origin of war in the state of nature, with the logical consequence that war, like inequality, is seen to be a result of society. In Montesquieu's thinking, war is less a human than a social phenomenon. From this it follows that if war and inequality are linked to the essence of society and not to the essence of man, the aim of politics will be, not to eliminate war and inequality, which are inseparable from collective life, but to mitigate or moderate them.

It is interesting to reflect on these two intellectual positions. Hobbes, who posited war as primeval and inherent in human nature, ended by authorizing the absolute power which alone is capable of keeping the peace. Montesquieu, on the other hand, considering war and inequality to be social phenomena, naturally had no incentive to do away with them altogether. By recognizing the social character of war, one relinquishes, as it were, the utopia of absolute peace. If war is a human phenomenon, we can dream of absolute peace. If war is a social phenomenon, we simply arrive at the ideal of moderation.

It would also be interesting to compare Montesquieu's position with that of Jean-Jacques Rousseau. We would find an opposition similar to the one between Montesquieu and Hobbes. Rousseau referred to a state of nature, conceived by human reason, which serves as a kind of criterion for society. By employing this criterion of what should be, he arrived at a conception of the absolute sovereignty of the people. Montesquieu, for whom the state of nature does not serve as a criterion, confined himself to stating that inequalities spring from society. He did not conclude from this that we must return to a natural equality, but simply that we must moderate as best we can the inequalities which derive from society itself.

Montesquieu's conception of the state of nature is not only indicative of the whole of his political philosophy; it is also the source of the two books he devoted to the laws of nations, i.e., Books IX and X. These two books are, in fact, oriented by assertions made at the beginning of the work, namely, those to be found in Chapter 3 of Book I:

> The law of nations is naturally founded on this principle, that different nations ought in time of peace to do one another all the good they can, and in time of war as little injury as possible, without prejudicing their real interests.
>
> The object of war is victory; that of victory is conquest; and that of conquest preservation. From this

and the preceding principle all those rules are derived which constitute the law of nations.

This passage is important because it shows that *The Spirit of the Laws* contains not only the causal scientific explanation of positive laws but also the analysis of the laws governing relations between collectivities in terms of the objective Montesquieu attributed to the law of nations. In other words, the goal toward which collectivities ought to tend may be determined by rational analysis.

Montesquieu's philosophy is neither the oversimplified determinism attributed to him by Auguste Comte, for example, nor a traditional philosophy of natural law. It is an original attempt to combine the two.

This accounts for the various interpretations that have been offered of his thought. I shall review several, taken from well-known authors.

The first of these is that of the German historian Friedrich Meinecke, who devoted a chapter of his classic work *Die Entstehung des Historismus,* to Montesquieu. Meinecke felt that Montesquieu's doctrine fluctuated between the rational universalism typical of so much eighteenth-century thought and the historical sense of particulars which was to flourish in the historical schools of the nineteenth century.

It is perfectly true that there are to be found in Montesquieu statements inspired by the philosophy of a rational and universal order alongside statements which emphasize the diversity of historical collectivities. But it is not so certain that Montesquieu's thought must be regarded as a clumsy compromise between these two inspirations, as a stage in the gradual discovery of pure "historicity." It can also be seen as a legitimate, if imperfect, attempt to combine two ways of thinking, neither of which can be entirely eliminated from the exercised human reason.

The second interpretation I have in mind is that of M. Althuser (*La Politique et l'Histoire*), who is a Marxist. Here the contradiction in Montesquieu is supposed to be simpler and more blatant, namely, between his original genius and his reactionary opinions. This interpretation has an element of truth in it. In the ideological conflicts of the eighteenth century, Montesquieu belonged to a group which, in certain respects, may be called reactionary. He advocated the return to, or restoration of, institutions which had existed in a more or less legendary past and which had disappeared. In the course of the eighteenth century, especially during the first half, the great controversy among French political writers focused on the theory of monarchy.[16] What was ideologically at issue, so to speak, was the position of the aristocracy, the nobility, in the monarchy. There were, broadly, two opposing schools. The Romanist school traced the absolute power of the French monarch to the sovereign empire of Rome, to which the king of France was held to be the heir. According to this view, the French king's claim to absolute power was justified by history. The second interpretation was the so-called Germanic one, which traced the privileged position of the French nobility to the conquest of the Gauls by the Franks and justified these privileges by historical ascription. (This controversy also resulted in doctrines which were developed in the following century and culminated in notions which were, properly speaking racist; for example, the doctrine that the nobility was Germanic and the people Gallo-Roman.) The right of conquest is today regarded as a poor justification for the continuation of a regime of inequality; but in the eighteenth century, it often passed for a legitimate basis of the social hierarchy.[17]

We need only to refer to the last three books of *The Spirit of the Laws* to discover that, in the controversy between these two schools, Montesquieu was on the side of the Germanic school—if always with modifications, reservations, and more subtlety than was commonly found among the intransigent theorists of the rights of the nobility. At the end of Chapter 6 of Book XI, the famous chapter on the English constitution, there occurs the statement that English liberty, based on the balance of power, was born "in the woods," i.e., in the forests of Germany. Montesquieu unquestionably was concerned with preserving the privileges of the nobility as an intermediary body of the realm.[18]

Montesquieu, it must be remembered, was in no sense a doctrinaire of equality, still less a doctrinaire of popular sovereignty. Since he related social inequality to the essence of the social order, he accommodated himself very easily to inequality. And if one believes with M. Althuser that popular sovereignty and equality are the political concepts which have prevailed through the revolutions of the nineteenth and twentieth centuries, through the French Revolution and the Russian Revolution—if one believes that history is moving in the direction of popular sovereignty and equality, then one is justified in calling Montesquieu a doctrinaire of the *ancien régime* and in saying that as such he was truly reactionary.

However, it seems to me that the question is a more complicated one. However history may judge the specific institutions to which he referred, his ultimate opinion was that the social order is essentially heterogeneous and that the condition of liberty is the balancing of social powers and the power of government by the *nobility*—the sense in which nobility embraces the best citizens of an equalitarian democracy, or even the militants of the Communist party in a regime of the Soviet type, as well as the nobility in a monarchy.

In other words, the essence of Montesquieu's political philosophy is liberalism: the goal of the political order is to insure the moderation of power by the balance of powers, by the equilibrium of people, nobility, and king in the French or the English monarchy, or the equilibrium of the people and privileged, plebs and patriciate, in the Roman republic. These are different examples of the same fundamental conception of a heterogeneous and hierarchical society in which the moderation of political power requires the balance of powers.

If this was the ultimate opinion of Montesquieu, it is by

no means proved that he was a reactionary. Undoubtedly he was a reactionary in terms of the French controversies of the eighteenth century. He neither foresaw nor desired the French Revolution. Perhaps he involuntarily prepared the way for it, for one never knows, either beforehand or after the fact, who prepared what in history; but he did not consciously desire the French Revolution. Insofar as one can foresee what a man would have done in circumstances other than those in which he lived, it is conceivable that Montesquieu, in case of absolute necessity, would have been a supporter of the Constituent Assembly. But he would very soon have gone over to the opposite side, and he would have had the choice, like the liberals of his kind, between emigration, the guillotine, or internal emigration on the fringes of the tumultuous vicissitudes of the Revolution.

But though a political reactionary, Montesquieu remains perhaps *the* representative of a style of thinking which is by no means outmoded or anachronistic. Whatever the structure of a society at a given period, it is always possible to think in the manner of Montesquieu, that is, to analyze the peculiar form of heterogeneity of a certain society and to seek, by the balance of powers, the guarantee of moderation and liberty.

The third interpretation I would refer to can be found in the short chapter on Montesquieu in Léon Brunschvicg's *Le Progrès de la Conscience dans la Philosophie Occidentale*, which asserts that Montesquieu's thought is fundamentally uneven and contradictory.[19]

Brunschvicg writes that, in a sense, Montesquieu has given us the masterpiece of pure sociology (by which he means analytical sociology). But Brunschvicg is inclined to think that, aside from this pure sociology, there is no system in Montesquieu. He quotes the lines which I have discussed, "Indeed, the rational world must be governed just as well as the material world," and emphasizes the paradox inherent in the rational world's possibility of violating the laws to which it is subject. Brunschvicg also

61

shows Montesquieu's fluctuation between Cartesian concepts (for example, that before one has drawn the circle, all the radii are already equal, just as justice and injustice exist before the positive laws), and the kind of quasi-empirical classification of types of government which derives from the Aristotelian tradition. Finally, he sees neither unity nor coherence in Montesquieu's thought and merely concludes that readers have nonetheless found in *The Spirit of the Laws* an implicit philosophy of progress, inspired by liberal values.

Personally, I feel that this judgment is harsh. It is true that there is no system in Montesquieu. But perhaps it is consistent with the spirit of a historical sociology that it have no system. I believe that an attentive reader will find that Montesquieu's thought, though exceedingly difficult to disentangle, is far from being as contradictory as it is often asserted to be.

Before I leave Montesquieu, I should like to answer two questions: Why is Montesquieu regarded not as a sociologist, but as a precursor of sociology? What is the justification for the fact that he is not placed among the founding fathers?

The first reason, the simplest one, which is easily guessed, is that the word *sociology* did not exist in Montesquieu's time and that this dreadful word, which has gradually become part of our culture, was invented by Auguste Comte.

The second, and far more profound, reason is that Montesquieu was not interested in modern society. Those we regard as the founders of sociology—Auguste Comte, say, or Marx—are interested in studying those qualities typical of modern society, i.e., typical of society conceived as basically industrial or capitalistic. Not only was Montesquieu not concerned with modern society, but the categories he used were to a large extent the categories of classical political philosophy. There is in *The Spirit of the Laws* neither the primacy of economics nor the primacy of society in relation to the state.

In one sense, Montesquieu was the last of the classical

philosophers. He was still a classical philosopher to the extent that he believed a society is essentially defined by its political regime and that his theory culminated in a conception of liberty. But in another sense, one can indeed say he was the first of the sociologists, for he reinterpreted classical political thought in terms of a total conception of society and he sought to explain all aspects of collectivities in a sociological mode.

Let us add that, in comparison with those who are today regarded as sociologists, Montesquieu is devoid of the ideology which constitutes almost a professional *sine qua non* among sociologists: the belief in progress. But that Montesquieu did not believe in progress in the sense in which Auguste Comte believed in it is not at all surprising; since for him the essence of societies was their political regime, he logically tended *not* to see a unilateral movement toward the good in the course of history. As Montesquieu—and many others before him—knew, *political* change consists of alternate advances and setbacks. Montesquieu assigned the central position in his analysis to politics, so he had to disregard the idea of progress, which naturally appears as soon as one considers economics or science. The economic philosophy of progress is to be found in Marx; the philosophy of human progress through science is to be found in Auguste Comte.

BIOGRAPHICAL CHRONOLOGY

1689	January 18. Birth of Charles-Louis de Secondat at the Château of Breda near Bordeaux.
1700–5	Studied under the Oratorians at Juilly.
1708–9	Studied law first in Bordeaux, and later in Paris.
1714	Admitted to the Bordeaux Parliament as Councilor.
1715	Married Jeanne de Lartigue.
1716	Elected to the Academy of Science in Bordeaux. He inherited from his uncle the presidency of the Bordeaux Parliament as well as all his wealth and the name of Montesquieu.

1717–21	Studied science and wrote articles on the echo, the kidneys, transparency, the weight of bodies, etc.
1721	Anonymous publication of *Persian Letters*. The book was an immediate success.
1722–25	Lived in Paris, where he led a gay, sophisticated life. He saw often the Duke of Bourbon's entourage, President Henault, and the Marquesa de Prie and frequented Madame Lambert's salon and the Entresol Club, where he read his *Dialogue Between Silla and Eucrates*.
1725	Anonymous publication of *The Temple of Gnide*. Back in Bordeaux, Montesquieu seceded from office and returned to Paris. He was to write later in his *Pensées:* "What I have always blamed myself for is how few positions there were in the Republic that I was truly well-fitted for. As for my presidential office, I had good intentions; I understood the questions in themselves well enough, but nothing of the procedure. Nevertheless I applied myself; but what disgusted me the most is that I saw in some idiots the very ability that escaped me." (O.C., t. I, p. 977.)
1728	Elected to the *Académie francaise*. Left for Germany, Austria, Switzerland, Italy, and Holland, from where he was eventually taken to England by Lord Chesterfield.
1729–30	Stayed in England.
1731	Returned to the Château of Breda, where he devoted his time to writing *The Spirit of the Laws*.
1734	Publication of *Considerations on the Causes of the Grandeur and Decadence of the Romans*.
1748	Publication of *The Spirit of the Laws*, edited anonymously in Geneva. It met with great success, but the book was more commented on than read.
1750	*Defense of the Spirit of the Laws*, an answer to attacks made by the Jesuits and the Jansenists.
1754	Wrote his *Essay on Taste* for the *Encyclopédie* at the request of D'Alembert. It was published in 1756.
1755	He died in Paris on February 10.

NOTES

1. One is reminded of J. M. Keynes's debatable sally in his preface to the French edition of *The General Theory:* "Montesquieu, the greatest French economist, who may justly be compared with Adam Smith, and who stands head and shoulders above the physiocrats by virtue of his perspicacity, the clarity of his ideas, and his common sense (qualities which every economist ought to possess)." (J. M. Keynes, *Théorie générale de l'emploi, de l'intérêt et de la monnaie,* translated by J. de Largentaye, Paris, Payot, 1953, p. 13.)

2. "There is this difference between the nature and principle of government, that the former is that by which it is constituted, the latter that by which it is made to act. One is its particular structure, and the other the human passions which set it in motion. Now, laws ought no less to relate to the principle than to the nature of each government." (*The Spirit of the Laws,* Book III, 1., p. 19.)

3. "For it is clear that in a monarchy, where he who commands the execution of the laws generally thinks himself above them, there is less need of virtue than in a popular government, where the person entrusted with the execution of the laws is sensible of his being subject to their direction. . . . When virtue is banished, ambition invades the minds of those who are disposed to receive it, and avarice possesses the whole community." (*Op. cit.,* Book III, 3., pp. 20–21.) "It is the nature of honor to aspire to preferments and titles. . . ." (*Op. cit.,* Book III, 7., p. 25.)

4. In fact, the fundamental distinction between republic and monarchy had already been made by Machiavelli: "All governments, all seigniories which have had and still have authority over men were and are either Republics or Principalities." (*The Prince,* Chapter 1.)

5. On this matter see F. T. H. Fletcher, *Montesquieu and English Politics,* London, 1939, and P. M. Spurlin, *Montesquieu in America 1760–1801,* Louisiana State University, 1940.

6. The texts of Locke with which Montesquieu worked are the *Two Treatises of Government: In the Former the False Principles and Foundation of Sir Robert Filmer and His Followers are Detected and Overthrown; The Latter is an Essay Concerning the True Origin, Extent and End of Civil Government,* published for the first time in London in 1690. The second of these two treatises, the *Essay Concerning the True Origin, Extent and End of Civil Government,* was translated into

French by David Mazel and published in Amsterdam by A. Wolfgang in 1691 under the title *Du Gouvernement civil, où l'on traite de l'origine, des fondements, de la Nature du Pouvoir et des fins des Sociétés politiques*. Numerous editions of the work in the Mazel translation appeared during the eighteenth century. (A new translation for a modern edition was made by J. L. Fyot under the title *Essai sur le pouvoir civil*, and published by the Bibliothèque de la Science politique, Paris, Presses Universitaires Françaises, 1953, with a preface by B. Mirkine-Guetzevitch and Marcel Prélot.)

Locke's theory of the powers and the relationship between them is set forth in Chapters XI through XIV of the *Essay on Civil Government*. In Chapter XII, Locke distinguishes three types of power: the legislative power, the executive power, and the federative power of the state. "The legislative power is that which has a right to direct how the force of the commonwealth shall be employed for preserving the community and the members of it." The executive power is "a power always in being which should see to the execution of the laws that are made, and remain in force." Hence it includes both administration and justice. In addition, "There is another power in every commonwealth which one may call natural, because it is that which answers to the power every man naturally had before he entered into society . . ." So that under this consideration the whole community is one body in the state of Nature in respect of all other states or persons out of its community.

"This, therefore, contains the power of war and peace, leagues and alliances, and all the transactions with all persons and communities without the commonwealth, and may be called federative, if any one pleases.

". . . These two powers, executive and federative, though they be really distinct in themselves, yet one comprehending the execution of the municipal laws of the society within its self upon all that are parts of it; the other the management of the security and interest of the public without, with all those that it may receive benefit or damage from, yet they are almost always united . . . it is almost impracticable . . . that the executive and federative power should be placed in persons that might act separately, whereby the force of the public would be under different commands, which would be apt sometime or other to cause disorder and ruin." (*Social Contract*, New York, Oxford University Press, 1962, pp. 85–87.)

7. This conception is not totally new. The interpretation of the Roman constitution in terms of the idea of the division and balance of powers and social forces is already found in the theory of mixed government of Polybius and Cicero. And these

writers more or less explicitly regarded this division and balance as a condition of freedom. But it is in Machiavelli that one reads statements that look forward to Montesquieu. "To those who condemn the quarrels between the Senate and the people I maintain that they are condemning what was the origin of liberty, and that they are much more impressed by the noise and confusion these quarrels caused on the public square than by the good effects they produced. In every Republic there are two parties, that of the nobles and that of the people, and all laws favorable to freedom are born only of their opposition." (*Discourse on the First Decade of Titus Livius,* Book I, Chapter 4.)

8. The theme of the separation of powers is one of the principal themes of the official constitutional doctrine of General de Gaulle. "All morality and all experience dictate that the public power—legislative, executive, and judiciary—be clearly separated and carefully balanced." (Speech in Bayeux, June 16, 1946.) "Let there be a government that is made to govern, which is allowed the time and opportunity to do so, which is not directed from its task, and which therefore merits the support of the governed. Let there be a parliament destined to represent the public will of the nation, to pass the laws, to control the executive, without wishing to step out of its role. Let government and parliament collaborate but remain separate in their responsibilities, and let no member of one be able to be at the same time a member of the other. Such is the balanced structure that power must assume . . . Let the judiciary authority be assured of its independence and remain the guardian of the individual freedom. Thus the competence, dignity, and impartiality of the State will be better guaranteed." (Speech at the Place de la République, September 4, 1958.) Let us note, however, that in the case of the Constitution of 1958, the executive arm can check the legislative arm more easily than the legislative the executive.

For interpretation by jurists of Montesquieu's theory of the separation of powers, see especially: L. Duguit, *Traité de Droit constitutionnel,* Vol. 1; R. Carre de Malberg, *Contribution à la théorie générale de l'Etat,* Paris, Sirey, Vol. 1, 1920, Vol. 2, 1922, especially Vol. 2, pp. 1–142; Charles Eisenmann, *"L'Esprit des lois" et la séparation des pouvoirs,* in *Mélanges Carré de Malberg,* Paris, 1933, pp. 190 ff. and "La pensée constitutionnelle de Montesquieu," in *Recueil Sirey du Bicentenaire de "L'Esprit des lois,"* Paris, 1952, pp. 133–60.

9. This interpretation is notably that of Louis Althusser in his book *Montesquieu, la politique et l'histoire,* Paris, Presses Universitaires Françaises, 1959, 120 pp.

10. And in Montesquieu's analysis of the republic, in spite of the essential idea that the nature of the republic is the equality of the citizens, there is a differentiation between the mass of the people and the elites.

11. Diderot; the Encyclopedists; and, above all, Voltaire, who defended Calas, Sirven, the Chevalier de La Barre, and other victims of the justice of his time and wrote an *Essai sur la probabilité en fait de justice* (1772), bear witness to the great interest aroused by penal questions in the eighteenth century. But the great moment in the penal controversy was the appearance in 1764 of the *Essay on Crimes and Punishments* by the Cesare Beccaria of Milan (1738–94). This work, written when its author was twenty-six years old, caused immediate comment throughout Europe, notably on the part of Father Morellet, Voltaire, and Diderot. Beccaria's treatise develops the idea that punishment should be based not on the principle of *restitutio juris* but on the relativistic and pragmatic principle of *punitur ne peccetur*. Furthermore, he radically criticizes the penal procedure—or lack of procedure—of his day and demands that punishments be proportionate to crimes. This work forms the basis of modern criminology and was the direct cause of subsequent penal reforms. Cf. M. T. Maestro, *Voltaire and Beccaria as Reformers of Criminal Law*, New York, 1942.

12. On this problem of the influences exerted on Montesquieu, one should refer to the work of J. Dedieu, one of the foremost authorities on Montesquieu: *Montesquieu et la tradition politique anglaise en France. Les sources anglaises de "L'Esprit des lois,"* Paris, Lecoffre, 1909, and *Montesquieu,* Paris, 1913.

13. "The more the physical causes incline mankind to inaction, the more the moral causes should estrange them from it." (*The Spirit of the Laws,* Book XIV, 5, p. 226.)

14. It would be unfair to reduce Montesquieu's economic analyses to this single error. Actually, Montesquieu presents a rather detailed and usually accurate picture of the factors that enter into the development of economics.

As an economist, he is not very systematic. He belongs neither to the mercantilist nor to the physiocratic school. But it is possible, as has recently been done, to see him as a sociologist who anticipated modern study of economic development precisely because he took into consideration the many factors that come into play. He analyzes peasant labor, the very foundation of the existence of collectivities. He distinguishes different systems of ownership, he tries to find the consequences of the different systems of ownership upon the number of workers and upon agricultural yield, and the relation between system of ownership and agricultural labor and volume of population. Next, he looks for the relation between volume of population and diversity of social classes. He outlines a theory that might be called the theory of luxury. There must be rich classes to maintain a market for useless objects, objects that do not correspond to an urgent necessity of existence. He considers the relation between internal commerce between the different social classes and the external commerce of the collectivity. He considers money; he traces the role of money in transactions within collectivities and between collectivities. Finally, he tries to discover to what extent a certain political regime does or does not promote economic prosperity.

This adds up to any analysis that is less partial and less schematic than those of economists in the strict sense of the term. Montesquieu's ambition was to construct a general sociology that would include economic theory properly speaking.

In this mode of analysis there is perpetual reciprocal influence between the different elements. Mode of ownership has an effect on the quality of agricultural labor, and the latter in turn has an effect on relations between social classes. The structure of social classes influences internal and external commerce. The central idea is the reciprocal, indefinite influence upon one another of the different sectors of the social whole.

15. According to Louis Althusser, in his book *Montesquieu, la politique et l'histoire,* the author of *The Spirit of the Laws* is responsible for a veritable theoretical revolution. This revolution "assumes that it is possible to apply a Newtonian category of law to matters of politics and history. It assumes that it is possible to derive from human institutions themselves principles with which to consider their diversity in terms of a unity and their fluidity in terms of a constancy: the law of their diversification, the law of their evolution. This law will no longer be an ideal order, but a relation immanent in phenomena. It will not be given in the intuition of essences, but derived from the facts themselves, without preconceived idea, by study and comparison, by trial and error" (page 26). But "the sociologist, unlike the physicist, does not deal with an object (the body) which obeys a simple determinism and follows a line from which it does not deviate, but with a very particular type of object: these men, who deviate even from the laws that they give themselves. What, then, can be said of men in their relation to their laws? That they change them, twist them, or violate them. But all this in no way changes the fact that one can derive from their behavior, whether submissive or rebellious, a law that they follow without knowing it, and even from their errors, its truth. To become discouraged from discovering the

laws of human behavior, one must be simple-minded enough to mistake the laws that they give themselves for the necessity that governs them: the fact is that their efforts, the aberration of their humors, the violation and changing of their laws are all simply part of their behavior. One must simply isolate the laws governing the violation of laws, or their changing . . . This attitude presupposes a very fruitful principle of method which consists in not mistaking the motives of human action for its driving force, nor the ends and reasons which men consciously propose for the real causes that animate them, which are usually unconscious." (Pp. 28–29.)

16. On this whole question of the ideological debate of the eighteenth century, reference should be made to Elie Carcassonne's thesis *Montesquieu et le problème de la Constitution française au XVIIIᵉ siècle,* Paris, 1927.

17. Louis Althusser summarizes the dispute as follows: "One idea dominated all the political literature of the eighteenth century: the idea that absolute monarchy was established against the nobility, and that the king enlisted the support of the commoners to balance the power of his feudal adversaries and reduce them to his mercy. The great quarrel of the Germanists and the Romanists over the origins of feudality and absolute monarchy developed against the background of this general conviction. On the one hand the Germanists (Saint-Simon, Boulainvilliers and Montesquieu, the latter being better informed and more subtle but just as firm) nostalgically evoke the era of primitive monarchy—a king elected by nobles, an equal among equals, as he was originally in the 'forests' of Germany—and contrast this with a monarchy that has become absolute—a king opposing and sacrificing the nobles, and choosing his representatives and allies from the commonality. On the other hand, the absolutist party of bourgeois inspiration, the Romanists (Father Dubos, author of a conspiracy against the nobility [*The Spirit of the Laws,* XXX, 10] and target of the last books of *The Spirit of the Laws*), and the Encyclopedists celebrate, either in Louis XIV or in the enlightened despot, the ideal of the prince who prefers the merits and rights of an industrious bourgeoisie to the outdated pretentions of the feudal lords." (*Op. cit.,* pp. 104–5.)

At the origin of Germanist traditionalism is an unpublished work by Father Le Laboureur commissioned March 13, 1664 by the peers of France to discover in history "the proofs of the rights and prerogatives attached to their rank." Le Laboureur, whose work was almost certainly known to Saint-Simon, believed he had traced the origin of nobility to the Frankish conquest and developed the theory of a nobility participating in the government with the king at the time of the assemblies of the Champ de Mars or the Champ de Mai. The Duc de Saint-Simon (1675–1755) in his bills of government, which were drawn up about 1715, and the Comte de Boulainvilliers (1658–1722) in his *Histoire de l'ancien gouvernement de la France* (1727), his *Mémoire présenté à Monseigneur le Duc d'Orléans Régent* (1727), and his *Essai sur la noblesse de France* (1732), developed this apology for the old monarchy—the "reign of the incomparable Charlemagne"—which shared its powers with the leuds according to the traditions of the Franks. Germanist feudalism persisted until the first half of the nineteenth century. Montlosier, in his *Traité de la monarchie française,* echoed in 1814 the themes of Boulainvilliers in order to defend "the historic rights of the nobility." Indeed, reactions to this kind of reasoning determined the vocations of a number of great historians of the generation of 1815, notably Augustin Thierry, whose early works (*Histoire véritable de Jacques Bonhomme* in 1820) took as their motto Sieyes' question, "Why should the Third Estate not send back to the forests of Franconia all those families who maintain the foolish pretense of being descended from the race of the conquerors?"

The Germanism of Le Laboureur and de Boulainvilliers was both "racist," in that it favored the rights of conquest, and liberal, in that it was hostile to absolute power and favorable to the parliamentary system. But the two elements were dissociable.

In its reference to Frankish traditions of liberty and to the assemblies of the forests of Germany, this politico-historical doctrine was not totally committed to the interests of the nobility. Father Mably, in his *Observations sur l'histoire de France* (1765), one of the books which undoubtedly had the greatest influence on the revolutionary generations, presented a version of this doctrine that justified the convocation of the General Estates and the political ambitions of the Third Estate. When, in 1815, Napoleon wanted to make his peace with the people and with liberty, he borrowed the idea of the Extraordinary Assembly of the Champ de Mai from Mably's book. Similarly, in the nineteenth century, Guizot, who has been called the historian of the legitimate rise of the bourgeoisie, was, like Mably, a confirmed Germanist (cf. the *Essais sur l'histoire de France* of 1823 or the 1828 *Leçons sur L'Histoire générale de la civilisation en Europe*).

Tocqueville and Gobineau are undoubtedly the last heirs of the Germanist ideology. With Tocqueville, feudalism takes the form of regret for the rise of monarchical absolutism, and reinforces the liberal convictions of his heart and the democratic convictions of his mind. With Gobineau who, via his uncle and Montlosier, drew his inspiration directly from the aristocratic doctrinaires of the eighteenth century, the liberal vein disappears in favor of racism (see the correspondence between Tocqueville and Gobineau in the *Oeuvres complètes* of Tocqueville, Vol. 9, Paris, Gallimard, 1959, and the preface by J. J. Chevallier).

18. This did not prevent him, however, from being clearheaded about his own milieu. His works are not lacking in sallies at the foibles and vices of the nobility and the courtiers. It is true that the satire against courtiers is more a satire against what the monarchy has made of the nobility than against the nobility itself or the nobility as it should be, that is, free and independent in its fortune. Thus "the corps of lackeys is more respectable in France than elsewhere, it is a seminary for great lords. It fills in the space between the other Estates" (*Lettres persanes,* No. 98, *Oeuvres complètes,* Vol. 1, p. 277) or again, "Nothing approaches the ignorance of the people of the Court of France, unless it is that of the ecclesiastics of Italy." (*Mes Pensées, Oeuvres complètes,* Vol. 1, p. 1315.)

19. Léon Brunschvicg, *Le Progrès de la Conscience dans la philosophie occidentale,* pp. 489–501.

Auguste Comte

I

As we have seen, Montesquieu the sociologist was first and foremost intensely aware of human and social diversity. For him, the problem was to create order in an apparent chaos. He succeeded in doing so by observing types of government or society, by listing the determinants which influence all collectivities, and perhaps, in the last analysis, by evolving several rational principles of universal validity, though these may be violated in some cases. Montesquieu started from diversity and arrived, not without difficulty, at human unity.

Auguste Comte may be considered as, first and foremost, the sociologist of human and social unity. Human history is a single entity in his eyes, and he extended this conception of unity to the point where his difficulty would ultimately be the opposite of Montesquieu's. He would have trouble rediscovering and accounting for diversity, because there is, according to his philosophy, only one type of society which is absolutely valid and all mankind must arrive at this exemplary type.

This being the case, it seems to me that the stages in Comte's philosophical evolution may be considered as representing the three ways in which the thesis of human unity is stated, developed, and justified. These three stages are marked by Comte's three major works. The first stage, from 1820 to 1826, is that of the *Opuscules,* the *sommaire appréciation de l'ensemble du passé moderne* and the *plan des travaux nécessaires pour reorganiser la société;* then the *Considérations sur le Pouvoir Spirituel* and the *Considérations sur la Science et les Savants*. The second stage consists of the lectures of the *Cours de Philosophie Positive* and the third stage, of the *Système de Politique Positive*.

In the first stage, in the *Opuscules* which Auguste Comte republished at the end of Volume IV of the *Système de Politique Positive*, he considered the society of his day. As I have remarked, sociologists generally choose as their point of departure an analysis of the historical period to which they belong. In this respect, Auguste Comte is typical; the *Opuscules* are a description and analysis of a moment in the history of European society.

According to Comte, a certain type of society is dying, another being born before his eyes. The dying type is characterized by two adjectives: theological and military. Medieval society was united by transcendent faith as expounded by the Catholic Church. Theological thinking is contemporaneous with the predominance of military activity, a predominance which is expressed by the fact that the highest rank is granted to warriors. The type being born is scientific and industrial. This society is scientific in the sense in which the moribund society was theological: the thinking typical of the modern age is that of scientists, just as the thinking typical of the past was that of theologians or priests. Scientists are replacing priests or theologians as the social category providing the intellectual and moral foundation of the social order. The scientists are inheriting the spiritual power of the priests. Spiritual power, according to Auguste Comte in his early *Opuscules*, is necessarily embodied in each age by those who provide the model for the predominant way of thinking and the ideas which serve as the basis of the social order. Moreover, just as the scientists are replacing the priests, the industrialists, in the broad sense of the word (i.e., in the all-inclusive sense of businessmen and managers and financiers), are replacing the warriors. Indeed, from the moment men think scientifically, the chief activity of collectivities ceases to be the war of man against man and becomes the struggle of man against nature, the systematic exploitation of natural resources.

The conclusion Comte drew from the analysis of the society in which he lived is that the basic condition of social reform is intellectual reform. It is not by the accidents of a revolution nor by violence that a society in crisis will be reorganized, but through a synthesis of the sciences and by the creation of positive politics.

Like many of his contemporaries, Comte believed that modern society was in crisis; as a result one social order was disappearing and another social order was being born.

The result of this analysis of the contemporary crisis in terms of the opposition between the theological-military social type and the scientific-industrial social type is that Comte the reformer was not a theorist of revolution *à la* Marx nor a theorist of liberalism *à la* Montesquieu or Tocqueville; he was a theorist of positive science, and especially of the social science he calls sociology.

In other words, from the interpretation of contemporary society follows the general orientation of thought. Montesquieu, as we have seen, was first of all the observer of the crisis of the French monarchy, a crisis which was one of the origins of his over-all conception. Comte was the observer of the contradiction between two social types—a contradiction which can be resolved only by the triumph of that social type which he calls scientific and industrial. This victory is inevitable, but it can be retarded or accelerated. The function of sociology, according to Comte, is to understand the necessary, indispensable, and inevitable course of history in such a way as to promote the realization of the new order.

In his second stage, i.e., in the *Cours de Philosophie Positive,* the ruling ideas have not changed, but the perspective is broadened. For in the *Opuscules* Comte was primarily an observer of contemporary societies and their history, i.e., the history of Europe, and it would be easy for a non-European to demonstrate that in his first *Opuscules* Auguste Comte naïvely regarded the history of Europe as synonymous with the history of the human race; or, to employ another formulation, he presupposed the exemplary character of European history. In his second stage, that is, in the *Cours de Philosophie Positive,* Comte gave more universal scope and deeper meaning to the idea of progress. In particular, he developed and corroborated the two basic themes which he had already expounded in the *Opuscules:* the law of the three stages of human evolution and the classification of the sciences.[1]

The law of the three stages consists in the assertion that the human mind passes through three phases. In the first, the mind explains phenomena by ascribing them to beings or forces comparable to man himself. In the second phase, that of metaphysics, the mind explains phenomena by invoking abstract entities like "nature." Finally, in the third phase, man is content to observe phenomena and to establish the regular links existing among them, whether at a given moment or in the course of time. He abandons the search for the final principle behind the facts and confines himself to establishing the laws that govern them.

But the transition from the theological age to the metaphysical age and thence to the positive age does not occur simultaneously for the various intellectual disciplines. In Comte's thinking, the law of the three stages has no precise meaning unless it is combined with the classification of the sciences. For it is the order in which the various sciences are ranked that reveals the order in which the intelligence becomes "positive."[2] The positive method was adopted sooner in mathematics, in physics, and in chemistry than in biology. There are reasons why positivism is slower to appear in disciplines relating to the most complex matters. The simpler the object of study, the easier it is to think positively. There are even certain phenomena in which observation follows automatically, and in these cases the intelligence has been positivist from the beginning.

The combination of the law of the three stages and the classification of the sciences eventually leads to Auguste Comte's basic formula: the method which has triumphed in mathematics, astronomy, physics, chemistry, and biology must eventually prevail in politics and culminate in the founding of a positive science of society, which is called sociology.

The point of this review by Comte of the various scientific disciplines was not just to demonstrate the need for creating sociology. It had a more specific aim: beginning with a certain science—namely, biology—there occurs a decisive reversal in methodology—a reversal which is to provide a foundation for the sociological concept of historical unity. Beginning with biology, the sciences are no longer analytic but necessarily and essentially synthetic.

These two terms, analytic and synthetic, have many meanings in Comte's terminology, as is evident from the following illustration. For Comte, the sciences of inorganic nature, physics and chemistry, are analytic in the sense that they establish laws among isolated phenomena. The separation of the phenomena or of relations is necessary and justifiable. On the other hand, in biology it is impossible to explain an organ or a function apart from the living creature as a whole. It is within and in relation to the whole organism that a particular biological fact assumes its meaning and finds its explanation. If we were arbitrarily and artificially to cut off a part of a living creature, we would have before us nothing but dead matter. One could say further that the living matter considered as such is an entity, a totality.

If we transpose the idea of the primacy of entity over element into sociology, we find that it is impossible to understand the state of a particular social phenomenon unless we restore it to its social context. We do not understand the state of religion, or the exact form assumed by the state in a particular society, unless we consider that society as a whole.

But this priority of entity over element does not apply merely to one moment arbitrarily cut off from the course of history. One understands the state of French society at the beginning of the nineteenth century only if one places this historical moment in the continuity of French history. One understands the Restoration only in terms of the Revolution, and the Revolution only in terms of the centuries of monarchical government. One understands the decline of the theological and military spirit only if one traces its origins in past centuries.

The priority of entity over element has, then, a second meaning, namely, that just as one element of social entity is understood only in terms of this very entity, so one moment in the historical evolution of this entity is meaningful only in terms of historical evolution as a whole.

But if one follows this line of thought, one encounters a very obvious difficulty: to understand one moment in the evolution of the French nation, one must refer to the entire history of the human race. The logical consequence of this principle of the priority of entity over element is the assertion that the object of sociology is the history of the human race. Auguste Comte was a man of logic, trained at the École Polytechnique; and since he had posited the priority of synthesis over analysis, he was obliged to conclude that the subject matter of the social science he wanted to establish—sociology—was the history of the human race regarded as a whole.

Here we see the inferiority or superiority (inferiority, in my opinion) of Comte as compared with Montesquieu. While Montesquieu began with the fact, which is diversity, Comte, with that intemperance in logic which is typical of great men and of some who are not so great, began with the unity of the human race and assigns sociology as its object of study nothing less than the history of that race.

Let me add that Auguste Comte, viewing sociology as a science in the manner of earlier sciences, did not hesitate to repeat the formula he had already used in the *Opuscules*, i.e., that just as there is no free will in mathematics or astronomy, there can be none in sociology. But, since scientists impose their verdict on the uneducated in mathematics and astronomy, they must logically impose their verdict in the same manner in sociology and politics. The obvious implication is that sociology can simultaneously determine what is, what will be, and what should be. But at the same time, what will and should be is justified as conforming to what the philosophers of the past would have called human nature or human destiny, or what Comte called simply the realization of the human and social order.

This brings us to the third stage in Comte's thought, in which this unity of human history is justified by a theory of both human nature and the nature of society.

The *Système de Politique Positive* is subsequent to Comte's attachment to Clotilde de Vaux, and his style and vocabulary had changed somewhat since the *Cours de Philosophie Positive*. But it is nevertheless true that the *Système de Politique Positive* still corresponds to a tendency in Comte's thinking already noticeable in his first, and especially in his second, stage. And if, as I believe, one can explain Comte's development by the desire to justify the idea of the unity of human history, it is natural that his last book should give this unity a philosophical foundation. For human history to be one, man must have a certain recognizable and definable nature at all times and in all places. Further, every society must admit of an essential order, whatever the diversity of social organizations. Lastly, this human nature and this social nature must be such that the major characteristics of historical evolution may be deduced from them. Now, in my opinion, the gist of the *Système de Politique Positive* can be explained by these three ideas.

Auguste Comte's theory of human nature comes under what he calls the *tableau cérébral*. Comte had certain ideas regarding the areas of the brain. But, setting aside certain peculiarities, this *tableau cérébral* is equivalent to an enumeration of the different activities characteristic of man. The basic social order recognizable in the diversity of institutions is described and analyzed in Volume II, the subject of which is "La Statique Social" (social statics). Finally, the *tableau cérébral* and social statics form the basis for Volume III of the *Système de Politique Positive,* which is devoted to dynamics. History as a whole tends toward the realization of the essential order of each society, analyzed in Volume II, and toward the achievement of what is best in human nature, described in the *tableau cérébral* of Volume I.

To summarize: The point of departure of Comte's thought is a contemplation of the internal contradiction of the society of his age, the contradiction between the theological and military type and the scientific and industrial type. He began with the analysis of a certain moment in history. Since this moment in history is characterized by the spread of scientific thinking and of industrial activity, Comte believed that the only way to put an end to the crisis is to create that system of scientific ideas which will govern the new social order, just as the system of theological ideas governed the social order of the past.

He then proceeded to the *Cours de Philosophie Positive,* in which he analyzed the whole of man's scientific achievement in order to determine the methods which have been applied in the various disciplines and the essential results that can serve in the creation of the science which is still lacking, namely, sociology.

But this sociology Comte wanted to create is not the prudent, modest, analytic sociology of Montesquieu, who endeavored to multiply the explanations for the innumerable institutions in all their diversity. Comte's science was meant to resolve the crisis of the modern world, to provide a system of scientific ideas which will preside over the reorganization of society.

For a science to preside over the reorganization of society, it must give results which are beyond doubt; it must furnish truths as incontestable as those of mathematics and astronomy. But these truths must be of a certain type. To be sure, Montesquieu's analytic sociology occasionally implies certain reforms; it frequently offers advice to legislators. But since it begins with the idea that the institutions of every society are conditioned by a multiplicity of factors, it eliminates the possibility of an institutional reality fundamentally different from the existing one. Now Comte wanted to be both a scientist and reformer. What science can be both positive in its pronouncements and useful to the reformer? Undoubtedly, a synthetic science as Comte conceived it, a science which would begin with the most general laws, the fundamental laws of human evolution, which would discover an over-all determinism which men could somehow put to use: in Comte's term, a modifiable fatality.

Comte's sociology began with what it is most interesting to know, i.e., ultimate truth. As for the details, they can be left to the historians, who were, in Comte's eyes, obscure drudges lost in meaningless erudition, objects of contempt to a man who instantaneously fathomed the underlying law of evolution.

In Montesquieu, politics or political organization apparently predominates. In Marx, as we shall see, economic organization predominates. Comte's doctrine is based on the idea that every society is united by means of the agreement of minds.[3] One might even say that society exists only to the extent that its members share the same beliefs. Thus, the different phases of human history are characterized by their way of thinking, and the present and final stage will be marked by the universal triumph of positive thought.

But after this point, having carried the notion of a single human history to its logical conclusion, Comte found himself forced to justify this unity, and he could justify it philosophically only by means of a conception of a coherent human nature and of a fundamental social order which is also constant.

Thus Comte's philosophy presupposed three major themes, which I shall have occasion to examine. The first is that industrial society, which in his eyes is equivalent to the society of Western Europe, will become the society of all mankind. And it has not been proved that Comte was wrong in thinking that certain aspects of European industrial society were destined to become universal. The industrial organization typical of European society is so much more efficient than any other form of organization that once a segment of humanity has found the "gimmick," if you will permit the expression, all the other segments of humanity must have it, too. Or, in more refined language, once the secret of the scientific organization of labor has been discovered, all men must lay hands on this condition of prosperity and power. Thus, the first sense in which human history is a unity, according to Comte, is the exemplary quality of industrial society.

The second theme is the double universality of scientific thinking. One can say that positive thought in mathematics, physics, or medicine has a vocation of universality, in the sense that all segments of the human race adopt this method of thinking once the results obtained by it have become visible. In this first version of universality, Auguste Comte was right enough. Western science has today become the science of all mankind, in mathematics, astronomy, physics, chemistry, and even, to a large extent, biology. But Comte had a second notion of universality: once positivism is introduced into astronomy or physics, it must be introduced into politics or religion as well. Now, this spread of the positive method is by no means obvious. Perhaps we are not doomed to imitate the method of mathematics or physics in sociology, ethics, or politics; at any rate, the question is still very much open.

The third of Auguste Comte's basic themes concerns the *Système de Politique Positive* and raises important questions: If human nature is fundamentally the same, if the social order is fundamentally the same, where does diversity come from? How is it to be explained and incorporated into his system?

II

COMTE'S early ideas were not particularly his own. The spirit of his age fostered the conviction that theological thinking was a thing of the past; that God was dead, if I may anticipate Nietzsche's formulation; that henceforth the human mind would be dominated by scientific thinking; that the feudal system or the monarchic structure was disappearing along with theology; that it was to be the scientists and industrialists who would dominate the society of our time.

All these themes were familiar in Comte's own day, but it is important to know which of them he selected and how he arranged them for the purpose of establishing his own interpretations of contemporary society.

The new fact which impressed all the observers of early nineteenth-century society—the basis of all their theories—was industry. They all believed that something novel was being born; this something was industry. But what was the originality of modern industry? What were its decisive features?

In my opinion, the characteristic features of industry, as observed by men of the early nineteenth century, were six.

(1) Industry constitutes a scientific organization of labor. Instead of proceeding according to custom, production is organized with a view to maximum output.

(2) As a result of the application of science to the organization of labor, man is engaged in a tremendous development of his wealth and resources.

(3) Industrial production implies concentrations of workers in factories and suburbs. A new social phenomenon is developing: the existence of the working masses.

(4) These concentrations of workers in industrial areas bring about an antagonism, whether latent or overt, between employees and employers, between proletariat and management or capitalists.

(5) Whereas wealth, as a result of the scientific character of labor, continues to increase, there is also an increase in crises of overproduction, which seem to create poverty in the midst of abundance. Crises of overproduction have an especially shocking quality, intolerable to the intelligence.

(6) The economic organization linked to the industrial and scientific aspect of labor is characterized by what is called free enterprise: profit-seeking on the part of management or merchants. Certain free-enterprise theorists even conclude that the condition of the increase of wealth is precisely unchecked profit-seeking and competition and that the less the state concerns itself with economic affairs, the more rapidly production and wealth will increase.

Contemporary interpretations varied according to the role played by each of these six features. Comte maintained that the first three are decisive. Industry is defined by the scientific character of labor, which results in the constant increase of wealth and the concentration of workers in factories. One might add that the concentration of workers in factories corresponds to the concentration of capital or the means of production in the hands of a few people.

The fourth point—the antagonism between workers and management—was secondary for Comte.[4] Antagonism is the result of bad organization of industrial society and can be corrected by reform. Similarly, Comte regarded the crises of overproduction as episodic and superficial phenomena.

As for free enterprise, Comte saw it not as the essence of the new society, but as a pathological element, a temporary aberration in the growth of an organization which is to be stabilized on principles other than those of unrestricted competition.

There is no need to remind you that, to the socialists, the two decisive features have been points (4) and (5). Marxist thought, like that of the pessimistic economists of the first half of the nineteenth century, develops from these two facts—the conflict between proletariat and management and the frequency of crises of overproduction—which are regarded as the inevitable consequence of capitalist anarchy. On the other hand, it is to point (6)—free enterprise—that the capitalist theoreticians assign the leading role and which they hold to be the decisive factor in economic progress.

Comte defined his own theory of industrial society by the criticisms he leveled at both the capitalist economists and the socialists. He offers a version of industrial society which is neither liberal nor socialist, but which might be defined as managerial—though this word is open to considerable misunderstanding and abuse (as in James Burnham's *The Managerial Revolution*).[5]

What are Comte's criticisms of the liberals or, more generally, of political economy? He reproaches the economists who speculate on "value"—who try to determine the functioning of the system in the abstract—for being metaphysicians. According to Comte, metaphysical thinking is abstract and conceptual, and such, in his opinion, was the thinking of the economists of his day[6]—who, moreover, committed the further error of examining economic phenomena apart from the social entity. (These two criticisms—the abstract character of political economy and the unjustifiable isolation of the economic aspect—have been taken over by the majority of French sociologists of the Durkheim school and have determined the semihostile attitude of the "sociologists" toward "economists" in French universities.) Finally, Comte reproached the capitalists for overestimating the effectiveness of the mechanisms of free enterprise or competition in increasing wealth.

Nevertheless the economists had one virtue in his eyes. This is the belief that, in the long run, private interests are in harmony. If the fundamental opposition between capitalists and socialists results from the fact that the former believe in the ultimate harmony of private interests and the latter believe in the inevitability of the class struggle, we may say that on this essential point Auguste Comte is on the side of the capitalists. He did not believe in a fundamental antagonism of interests between proletariat and management. There may be a temporary and superficial rivalry in the distribution of wealth; but, unlike the capitalist economists, Auguste Comte believed that the increase of wealth is (by definition, so to speak) consistent with the interests of all and that the basic law of industrial society is this increase of wealth and thereby the ultimate harmony of interests.

In relation to the economists who regard free enterprise and competition as the primary mechanisms for the increase of wealth, Auguste Comte belongs to the school of those I should call polytechnician-managers. He is, indeed, the symbolic patron of this school.

The polytechnician-manager is hostile to socialism or, more accurately, to those whom Comte generally called communists, since they were opposed to private ownership. And Comte in particular believed in the virtues, not certainly of competition, but of private ownership—and even, more curiously, the private ownership of concentrated wealth.

Comte justified the concentration of capital and of the means of production, a concentration which did not seem to him to be inconsistent with private ownership. Why?

The first reason, to borrow Comte's own idiom, is that concentration is inevitable; and if concentration is inevitable, then—by virtue of that providential optimism which is characteristic of Comte's philosophy of history—it must be salutary. Concentration is consistent with the basic tendency which may be seen in the course of human history. Material civilization can advance only if each generation produces more than it needs to survive and consequently transmits to the following generation a greater accumulation of capital than it received. This capitalization of the means of production is characteristic of the development of material civilization, and this capitalization entails concentration.

Comte was completely deaf to the argument that the

enormity of concentrated capital should imply the public character of ownership. Why did the concentration of the means of production not lead him to deduce their inevitable nationalizations? I think the reason is that Comte was more or less indifferent to the distinction between private and public ownership, because in his eyes authority, whether economic or political, is always personal. In every society, there are a few men who command. Now, one of the motives, conscious or unconscious, for the claim of public ownership is the belief, founded or unfounded, that the substitution of one regime of ownership for another would change the structure of the social order. Comte was skeptical on this point; in his eyes the rich are those who will always possess that share of power which of necessity accompanies great concentrations of wealth. In every social order, there are men who command; in every social order, there are concentrations of wealth; and it is desirable, in Comte's eyes, that the rulers be the men who possess these concentrations of wealth.

But—and this brings us to the second aspect of Comte's thought on this matter—such personal ownership must be purged, as it were, of its arbitrarily personal quality, since those whom he called the patricians, the temporal leaders, industrialists, bankers, etc., must regard their role as a social duty. Private ownership is necessary, inevitable, and indispensable; but it is tolerable only when conceived no longer as the right to use and abuse, but as the exercise of a collective function by the few whom fate or merit has singled out.[7]

This conception, it will be noted, is not so far from a certain kind of "Catholic socialism." But to this theory of private ownership Comte added another idea which assumes special importance in his last books: the idea of the secondary character of the temporal hierarchy.

Comte was all the more willing to accept the industrialists' concentration of wealth and authority since, for him, human existence is not defined exclusively by the position men occupy in the economic and social hierarchy. Beyond the temporal order that is subject to the law of power is a spiritual order which is an order of moral worth. The worker at the bottom of the temporal hierarchy may have a high rank in the spiritual hierarchy if his merits, his devotion to the collectivity, outweigh those of his superiors in the hierarchy.

What is this spiritual order? It is not a transcendent order, as the Catholic religion has conceived it. It is not an otherworldly order; it is not an order of eternal life; it is an order of this world, but one which substitutes an order of moral worth for the hierarchy of wealth and power. The supreme goal of every man should not be primacy in the hierarchy of power, but primacy in the hierarchy of merit.

In other words, if Comte's ambitions for economic reform were limited, it is because the central idea of his theory of industrial society is that the latter can endure only if it is regulated, tempered, and transfigured by a spiritual power.

Such are the major themes of Comte's analysis of industrial society. Now, this analysis has played an almost negligible role in the development of economic and social thought, at least in Europe. The Comtist conception of industrial society has remained a kind of curiosity on the fringes of the rivalry among doctrines because none of the political parties, either on the right or on the left, has acknowledged it, with the exception of a few individuals (of whom some, for that matter, belonged to the extreme right and others to the extreme left).

Among the French writers of this century, two have called themselves followers of Auguste Comte. One was Charles Maurras, the theorist of the monarchy, and the other was Alain, the theorist of radicalism. Both called themselves positivists for somewhat different reasons. Maurras was a positivist because he regarded Comte as the theorist of organization, of authority, and of a return to spiritual power.[8] Alain was a positivist because for him Comte's essential idea was the devaluation of the temporal hierarchy. He would readily have glossed over Comte's thought in the light of Kant's formula, "When I meet a powerful man, my body bows, but not my mind or my soul." These are not Kant's exact words, but this is the Kantian idea Alain frequently[9] quoted.

Why did Comte's conception remain outside the main stream of modern social thought? I think the question deserves to be asked, because in a certain sense Comte's theory is closer to those which seem to be coming into vogue today than are many other nineteenth-century doctrines. In fact, all the theories which today emphasize the similarity of a great many institutions on both sides of the iron curtain, which reject the importance of the mechanisms of economic competition—all those theories which attempt to isolate the essential characteristics of industrial civilization —could rightly claim to derive from Auguste Comte. He is, as it were, the theorist of industrial society, this side of or outside of the doctrinal disputes between capitalists and socialists, between theorists of free enterprise and theorists of planning.

The fundamental themes of free labor, of the application of science to industry, of the predominance of organization, are fairly typical of the modern conception of industrial society. Then why is Auguste Comte forgotten or ignored? The reasons, I think, are these.

I have discussed the leading ideas of Comte's theory, and not the detailed description of modern industrial society which he has given in the *Système de Politique Positive*. Now, if his leading ideas are profound, his detailed description of industrial society is often liable to ridicule. For Comte attempted to describe in detail the structure of the temporal hierarchy: the exact role of the temporal leaders, the industrialists, the bankers. He tried to show why those who perform the most general functions would have the most authority, would be situated highest in the hierarchy. He tried to specify the ideal number of men in each city and the number of patricians. He tried to explain how wealth would be transmitted. In short, he made an exact diagram of his dreams, or of the dreams each of us may invent in those moments when he takes himself for God.

Moreover, Comte's conception of industrial society is linked to the belief that wars have become an anachronism.[10] For history to prove Auguste Comte right as a philosopher of history, industrial society must be a peaceful society. But there is no question that between 1840 and 1945, history failed Auguste Comte. There have been, during the first half of the twentieth century, several wars, and wars of exceptional violence, which have disillusioned the loyal disciples of Auguste Comte.[11] Comte wrote that war would disappear from the *avant-garde* of the human race, that is, from Western Europe. But it was precisely Western Europe that has been the source and center of the wars of the twentieth century.

Also, according to Comte, the Western minority which, by good fortune, was at the head of humanity's progress should not conquer people of other races to impose its industrial society upon them. He had explained, using excellent arguments—that is, arguments which seemed excellent to him and which seem so to us as a result of the wisdom of hindsight—that the Westerners must not conquer Africa and Asia and that if they made the mistake of propagating their culture at gunpoint, the results would be disastrous both for them and for the others. Thus, Comte was right in his prediction of the consequences, but wrong in his prediction of the fact.[12]

Comte predicted peace because it was his opinion that war no longer had a function in an industrial society. War had been necessary in the past to force men, who were naturally lazy and anarchical, to some regular employment. It had been necessary for the creation of large states and a unified Roman empire, throughout which Christianity would spread and from which positivism would ultimately emerge. Thus, war had served a twofold historical purpose: the apprenticeship to work and the creation of large states. But in the nineteenth century, war no longer had a function. Society was characterized by the absence of a military class and the primacy of labor and the values of labor.[13] In the past, conquest might have been a legitimate means —or at any rate a rational means—for those profiting from it to augment their wealth. But as a method of acquisition, it was obsolete. In an age when wealth depended on the scientific organization of labor, the taking of booty was devoid of meaning. The transmission of property was henceforth effected by gift and by trade, and according to Comte, gifts were to play an increasing role and were even to reduce the role of trade to a certain extent.[14]

Comte's philosophy laid the greatest emphasis on the reform of the temporal organization by the spiritual power. The latter was to be the concern of scientists and philosophers, who were to replace the priests. The spiritual power was to rule men's feelings, rally them to a common task, sanction the rights of rulers, temper the despotism or egotism of the powerful. Now, it is probably on this point that history has disappointed Comte's disciples most harshly. Even if the temporal organization of industrial society is similar to what Auguste Comte imagined, the spiritual power of the philosophers and scientists is not yet born. What spiritual power there is, is exercised either by the churches of the past or by ideologists of some economic and social doctrines whom Auguste Comte would not have regarded as true scientists or philosophers.

In other words, insofar as men who claim to be scientific observers of the social order exercise a spiritual power—for example, in the Soviet Union—they emphasize, not the features common to all industrial societies, but a particular doctrine of the organization of industrial society. None of the various parties pays much deference to the man who underestimated the ideological conflicts by which the European societies have lived and for which so many millions have died.

Auguste Comte would have liked to see a spiritual power exercised by the interpreters of the social organization, who at the same time would have reduced the importance of the temporal hierarchy. This sort of spiritual power has never existed and does not exist today. Probably men always prefer what divides them to what unites them. Probably each society is compelled to emphasize its own individuality rather than the traits which it shares with all societies. Probably, too, societies are not yet sufficiently convinced of the merits which Comte attributed to industrial society in general.

According to him, the scientific organization of the industrial order resulted in assigning each man a position commensurate with his merits. Comte believed that if, in the past, age or birth had given certain men the highest rank, henceforth in the industrial society it was merit which would determine every man's position. And here is another of the ironies of history: there appeared some years ago in England a satirical book about a regime called The Meritocracy.[15] This was precisely Comte's idea of what industrial society ought to be. We can be sure that the author of the book never read Auguste Comte; nor would Comte have recognized his own dreams in the author's ironic description of such a regime. The irony arises from the fact that if everyone has a position commensurate with his merits, those who occupy the lowest positions can no longer blame fate or injustice. If all men were convinced that the social order was fair, then the latter would, in a certain sense and for certain persons, be intolerable—unless men were at the same time convinced of Auguste Comte's thesis that the hierarchy of intelligence is as nothing compared with the only hierarchy that counts: the hierarchy of merit and of the heart. But men are not easily persuaded that the temporal order is of no account.

III

THE SECOND STAGE in Comte's intellectual itinerary, which is also the second version of his philosophy, can be studied mainly in the light of the *Cours de Philosophie Positive,* particularly Volume IV, which expounds the idea of the new science called sociology.

First of all, who are the authors whom Auguste Comte quoted, whom he acknowledged as his precursors? Aside from Aristotle, with whom we shall deal later on, let us consider three: Montesquieu, Condorcet, and Bossuet. The juxtaposition of these three names will guide us, I believe, to some of the fundamental themes of Comte's thought in the field of sociology.

Montesquieu had one merit in Comte's eyes: the outstanding one of having posited the principle of determinism as regards historical and social phenomena. Comte offered a simplified interpretation of Montesquieu's thought; to him, Montesquieu's central idea is the one contained in the famous statement in Book I of *The Spirit of the Laws:* "Laws are the necessary relations arising from the nature of things." Comte found in this statement the principle of determinism as applied both to the diversity of social phenomena and to the evolution of societies.

Comte added that Montesquieu lacked another crucial idea: that of progress. Auguste Comte discovered the idea of progress in Condorcet, in the famous book *Tableau des Progrès Historiques de l'Esprit Humain*.[16] According to Condorcet, we can discover in the past a certain number of phases through which the human mind has passed. The number of these phases is fixed, and their succession is ineluctable. Comte borrowed from Condorcet the idea of progress of the human mind that is also an evolution of human society.

If we combine these two themes—Montesquieu's theme of determinism and Condorcet's theme of necessary sequential stages in the progress of the human mind—we arrive at Comte's central idea, which may be expressed in the following manner: social phenomena are subject to a strict determinism which operates in the form of an inevitable evolution of human societies—an evolution which is itself governed by the progress of the human mind.

Why have I added Bossuet's name to those of Montesquieu and Condorcet? The answer is that if you imagine the historical process as being determined in this way, you end with a secular view of history which is surprisingly analogous to Bossuet's religious, providentialist view. Thus, Comte wrote:

It is surely to our great Bossuet that we shall ever be indebted for the first important attempt of the human mind to contemplate from a sufficiently elevated perspective the whole of the history of society. Undoubtedly, the easy but illusory resources in the arsenal of any theological philosophy—ways of establishing among human events a certain apparent continuity—by no means allow us to employ today, in the construction of the true science of social development, the kinds of explanations which are inevitably preponderant in, and utterly irresistible to, such a philosophy. But this admirable composition—in which the spirit of universality, indispensable to any such conception, is so energetically maintained, insofar as the nature of the method employed permitted—will nonetheless ever remain an impressive model, always eminently suited to indicate in the clearest terms a general goal which our intelligence must never cease to set itself, a final result of all our historical analysis, I mean the rational co-ordination of the fundamental sequence of the various events of human history according to a single design, at once more real and more extensive than that conceived by Bossuet.

This last clause, "rational co-ordination of the fundamental sequence of the various events of human history according to a single design," is, as it were, the key to Auguste Comte's conception of sociology. When I remarked earlier that Auguste Comte was the sociologist of human unity, I neither exaggerated nor simplified his fundamental theme. Comte was concerned precisely with "the fundamental sequence of the various events of human history according to a single design." He was concerned with reducing the seemingly infinite variety of human societies in time and space to a fundamental sequence: the development of the human race, and to a single design: the culmination in an ultimate state of the human race and of the human mind. So we see how the man who is regarded as the founder of positivism can also be described as the last disciple of Christian or theological providentialism. We see, too, how the transition is made from the interpretation of human history in terms of divine providence to the interpretation of human history in terms of general laws.

Since this theme is the heart of Comte's thought—even in the *Cours de Philosophie Positive,* where it assumes its most scientific form—the purpose of my discussion will be to reveal the fluctuation, as it were, between the scientific idea and the providential one; or, better, to show how easy it is to shift from a certain conception of science to a new version of providence.

The single design—to repeat the expression used by Comte with reference to Bossuet—the single design of history as conceived by Comte is the progress of the human mind; and if the progress of the human mind gives unity to the entire history of society, it is because, according to Comte, the same way of thinking must prevail in all realms of thought. Comte observed that the positivist method was indispensable in the sciences of his day; and from this he concluded that the positivist method—based on observation, experimentation, and the establishment of general laws—must be extended to areas which were still entrusted to theology, i.e., to a method which claims to discover the underlying causes of phenomena and to locate final causes in transcendent beings.

According to Comte, then, there is a way of thinking—or a method which has universal validity—in politics as well as in astronomy. From this Comte concluded that because there is no free will in astronomy, there should be none in politics. But, for the moment, I should like to discuss an idea complementary to the preceding one, although it may seem to contradict it.

Comte declared that there cannot be true unity in a society when there is no body of ruling ideas adopted by all the different members of the collectivity. Society is heterogeneous, chaotic, in crisis, when there are opposing ways of thinking, when ruling ideas taken from incompatible philosophies are juxtaposed.

From this statement, one might seemingly conclude that in the past those societies which were not in crisis must have possessed a coherent body of ideas. But this conclusion would be only partly true, for the various sciences reach "the positive stage" at different moments in history. The sciences which reach it first are those which come first in Comte's classification of the sciences: mathematics, astronomy, physics, chemistry, biology. But this means that in all ages there have been sciences which were already partly positive, while other intellectual disciplines were still fetishistic or theological. Unity of thought, Comte's ultimate aim, has never yet been fully realized in the course of history.

In other words, one of the mechanisms of the movement of history is precisely the incoherence, at each stage of history, of various ways of thinking. In the last analysis, there has only been one period prior to positivism when there existed a true intellectual coherence, and that was pure fetishism. Fetishism is the immediate and spontaneous mode of thought of the human mind. It consists in animating all things, living or otherwise, in assuming things and creatures to be similar to man and to the mind of man. The mind will not rediscover a true coherence until the final phase, when positivism will have extended to all intellectual disciplines, including politics and ethics.

Between fetishism and positivism, there has been simultaneous diversity of methods of thinking, and it is this diversity that has prevented human history, at any one moment, from coming to an end.

It is true that at the start of his career Comte began with the notion that there could not be two different philosophies in any one society; but the development of his thought compelled him to admit that plurality of philosophies has been the dominant fact almost continually throughout history. The goal of social development is to lead human thought to the unity for which it is destined and which can be realized only in two ways; by immediate fetishism or ultimate positivism. One either explains all things by assuming them to be animated by human consciousness or abandons every theological and metaphysical explanation and confines oneself to establishing positive laws.

This raises, it seems to me, a basic question: Why is there such a thing as history? If the final and perfected state of human intelligence is to be positivism, why has humanity had to pass through so many successive stages? Why has it been necessary to wait for so many millennia for the appearance of the one man—namely, Auguste Comte—who has finally understood what the human mind should be?

The answer is that positivism cannot be a spontaneous philosophy; it can be only a latter-day philosophy. Indeed, it consists, for man, in recognizing the order which is outside himself, in admitting his inability to give a final explanation of it, and in confining himself to deciphering that order. But to discover this external order of nature requires time. Positivism consists in observing phenomena, in analyzing them, in discovering the laws governing the relations among them. But it is impossible through observation and analysis to discover this external order all at once. Before philosophizing, man must first live. In the earliest phase of the human adventure it was possible, at best, to explain a few simple phenomena in a scientific manner.[17] But a positivist philosophy—a philosophy of observation, experimentation, analysis, and determinism—could not be based on these few phenomena. In the initial phase of history, man needed another philosophy, which Comte at first calls theological and then fetishist. This philosophy enabled man to survive; it consoled him by describing the world as peopled with creatures like himself and therefore intelligible and benevolent. Fetishist or theological philosophy provides the human race with a temporary synthesis, both intellectually and morally valid: intellectually because it gives assurance of the intelligibility of the external world of nature, morally because it gives man confidence in himself and in his ability to overcome obstacles.

But the answer to the question, why is history necessary, raises another: Why must history necessarily continue to the very end?

Comte's answer seems to me to be something like this: Because certain phenomena have been explained scientifically and positively from the very beginning, a halt in the progress of the human mind is difficult to imagine. The contradiction between partial positivism and the fetishist, theological synthesis torments humanity, as it were, and in the last analysis prevents the human mind from stopping at any stage previous to the final one of universal positivism. But it must be added that, according to Comte himself, various segments of humanity have in fact been able to establish a temporary synthesis in one or another of their phases. And at the end of his life, Comte even decided that certain peoples could leap from the initial synthesis of fetishism to the final synthesis of positivism, skipping the intermediate phases which he describes in his social dynamics.

Because history is fundamentally the history of the progress of the human mind, another question arises: What is the relation between the movement of the intelligence and the other aspects of society, the other human activities? It is true that Comte himself does not raise the problem in the terms I am using. At no point does he ask himself what is the relation between the progress of the human intelligence and transformations of the economy, of war, or of politics. But it is easy to find the solution to the problem in his analyses.

In the *Cours de Philosophie Positive,* Comte states that history as a whole is essentially the development of the human mind. Here is a typical passage which occurs in Volume IV.

The significant part of this development, the part that has had the greatest influence on the general advancement, undoubtedly consists of the continual development of the scientific mind, from the early efforts of Thales and Pythagoras, to those of a Lagrange and a Bichat. Now, no enlightened person today could doubt that in this long succession of efforts and discoveries, human intelligence has always followed a clearly determined course, the exact foreknowledge of which would somehow have enabled a sufficiently informed intelligence to predict, before their more or less imminent realization, the basic advances reserved for each age.[18]

Obviously, Comte left little to chance or accidents. The crucial advances in the history of the human mind could have been foreseen by a superior intelligence because they answered a discernible need. But this does not mean that the movement of the intelligence *determines* the transformation of other social phenomena. Comte did not believe in the determination of the social entity by intelligence any more than Montesquieu believed in the determination of the social entity by the character of the political regime. The movement of history, for both men, is effected by action and reactions between the various segments of the total social reality.[19] If we consider Comte's social dynamics, whether in Part V of the *Cours de Philosophie Positive* or in Part III of the *Système de Politique Positive*, we see how the transition from one stage to the next is effected primarily by means of the opposition between the different segments of the society. Depending on circumstances, the cause which provokes disintegration of a social entity and the advent of the next stage may be in politics, in economics, or in intelligence.

Nevertheless, the primacy of the development of the intelligence is valid for Comte because (*a*) the major stages in the history of the race are determined by the dominant way of thinking, (*b*) the final stage is that of universal positivism, and (*c*) the chief instrument of human development is the constant criticism which nascent and maturing positivism brings to bear on the temporary synthesis of fetishism and theology.

Here we return to our central theme: that human history must be regarded as the history of a single people. (The expression is Auguste Comte's own, and it is intelligible enough.) If history were the history of religion, then in positing the unity of human history it would be necessary to posit nothing less than a "universalisable" religion. But if history is the history of intelligence, it is relatively easy to imagine a way of thinking valid for all men. To illustrate this idea with a simple example, let us say that today's mathematics seem to us to be true for all men of all races. (I am aware that this proposition is not self-evident; Spengler declares that there is Greek mathematics, just as there is modern mathematics. But Spengler himself advances this proposition only in a very special sense. He believes that the mathematical way of thinking is influenced by the particular style of a culture; I do not think he would have denied that mathematical theorems are universally true.[20] Insofar as all the propositions of all the intellectual disciplines can be reduced to theorems, it is conceivable that today's philosophy might be valid for all men and that, by the same token, history, conceived as the history of a single people.

But, granting that history is the history of a single people, that its stages are necessary, and that there is an inevitable progression toward a certain goal, what then accounts for the different histories of the different portions of humanity? As I have said, Montesquieu has trouble rescuing unity from our comprehensive theories; the problem for Comte is to rescue diversity.

In Part IV of the *Cours de Philosophie Positive*, we find Comte's solution to the problem of diversity. The factors of variation, he states, are three in number: race, climate, and political action.[21]

Comte, especially in the *Système de Politique Positive*, interpreted the diversity of the races of man by attributing to each of them a propensity toward a certain attitude—

more precisely, by ascribing to each of them the predominance of certain dispositions. For example, the black race was characterized primarily by the predominance of affectivity (which, at the end of his career, seemed to Comte to constitute a moral superiority). In other words, the different segments of humanity have not developed in the same way, because they did not have the same endowment. But these diversities develop against the background of a common nature. I shall discuss the Comtist conception of human nature later on.

Climate, for Comte, is a word which designates the ensemble of natural conditions in which each portion of humanity is to be found. Each society has had to overcome obstacles of varying difficulty; each has endured more or less favorable geographical circumstances; and this enables us to account up to a point for the diversity of their development.[22]

Finally, there remains a third factor of variation which I should like to emphasize: political action. Here we meet once again with Comte's providentialism, for our author is determined to strip politicians and social reformers alike of the illusion that an individual, however great, is able to alter substantially the necessary course of history. To be sure, Comte does not deny that it depends on circumstances, coincidences, or great men to determine how swiftly the necessary evolution takes place or how dearly the result—inevitable in any event—will cost. But if we consider the case of Napoleon, for example, we shall have no trouble discovering the limits of the potential effectiveness of great men.

According to Comte, Napoleon had failed to understand the spirit of his age or, as we would say today, the direction of history. He made a vain attempt to restore the military regime: he hurled France into the conquest of Europe, multiplied conflicts, aroused the peoples of Europe against the French Revolution; and, in the end, nothing came of this temporary aberration. The characteristic greatness of an age depends on the sum of circumstances; the sovereign who makes the mistake of failing to understand the nature of his age leaves no final mark.[23]

This theory—that individuals are incapable of changing the course of history—is part of a larger criticism of social reformers, utopians, or revolutionaries, those who think that either by designing a new society or by using violence on the old one the course of history can be reversed. It is true that fatality is increasingly modifiable, as one turns from the astronomical order to the historical one. As a result of sociology, which discovers the order essential to human history, we may be able to shorten the delays and reduce the cost of the advent of positivism. But by virtue of his theory of the inexorable course of history, Comte is hostile to both the illusions of great men and the utopias of the reformers. The sociology which he offers is the study of the laws of historical development. This sociology claims to be historical because, according to Comte, it will be based on observation and comparison, hence on methods comparable to those used by other sciences (biology, for example); but observation and comparison will somehow be controlled by the dominant themes of Comte's thought, by his conception of statics and dynamics. This means, in the case of statics, to comprehend the structure of a given society; in the case of dynamics, to comprehend the broad outlines of history. In both cases, it is to subordinate partial observations to the prior comprehension of the whole.

I shall devote the rest of this chapter to a brief analysis of the two concepts of statics and dynamics, which will carry us into the subject of the next chapter, namely, the analysis of human nature and social nature.

Statics and dynamics are the two basic categories of Auguste Comte's sociology and are related to the philosophy whose broad outlines I have sketched. Statics consists essentially in examining, in analyzing what Comte calls the social *consensus*. A society is comparable to a living organism. It is impossible to study the functioning of an organ without placing it in the context of the living creature. By the same token, it is impossible to study politics or the state without placing them in the context of the society at a given moment. Social statics thus consists, on the one hand, of the analysis—the anatomical analysis, if you will—of the society's structure at a given moment; and, on the other hand, of the analysis of the element or elements which at a given moment determine the consensus, which makes the collection of individuals into a collectivity, the plurality of institutions into a unity.

But if statics is the study of the consensus, it leads us to ask what are the essential organs of every society—and hence to go beyond the diversity of historical societies in order to discover the principles of every social order. Social statics begins as a simple positive analysis of the anatomy of various societies, of the bonds of mutual solidarity among the institutions of a particular collectivity, only to end, in Part II of the *Système de Politique Positive*, with the analysis of the essential order of every human collectivity.

As for dynamics, at the outset it consists merely of the description of the successive stages through which human societies pass. But since we start from the entity, since we know that the development of human societies and of the human mind is governed by laws, and since the whole of the past forms a unity, social dynamics does not resemble the history written by historians gathering facts or observing the sequence of institutions. Social dynamics will retrace the successive and necessary stages of the development of the human mind and of human societies. As social statics has revealed the essential order of every human society, social dynamics will ultimately retrace the vicissitudes through which this fundamental order has passed before arriving at the final goal of positivism.

If there is an essential order of all societies, dynamics must be subordinated to statics. By starting from the anatomy of every human society, we shall understand what history is. Then, instead of the terms *statics* and *dynamics*, it will be better to use the terms *order* and *progress*. One is reminded of the motto which appeared on the flag of positivism (and of Brazil!): "Progress is the development of order."[24]

At the outset, statics and dynamics are simply the study of coexistence on the one hand and succession on the other. At the conclusion, they are the study of the essential human and social order, of its transformations and fulfillment. But the transition from the apparently scientific terms—statics and dynamics—to the obviously philosophical terms—order and progress—is necessary as a result of Comte's two ideas: (1) the primacy of the social entity and of the laws applying to the entity; (2) the identification of the inevitable movement of history into a kind of providentialism. By means of a so-called scientific necessity, Comte rediscovered the equivalent of Bossuet's general design.

IV

IT IS IN THE *Système de Politique Positive* that the Comtist conception of statics finds its full development. All of Part II of the *Système* is given over to social statics. There is, of course, an outline of statics in the *Cours,* but this comprises only one chapter, and the ideas are merely suggested. Moreover, there are differences of detail between the ideas in the *Cours* and those in the *Système;* but in this summary I shall ignore the differences in order to examine social statics as conceived by Comte at the period he wrote the *Système de Politique Positive.*

This concept of statics can be divided logically into two parts: the study of human nature, on the one hand, and the study of social nature, on the other; or, again, the structure of human nature and the structure of social nature.

Auguste Comte expounded his ideas of human nature in what he called the *tableau cérébral,* which he offered as a scientific analysis of where, in the human brain, the anatomical equivalents of the various human dispositions are located. But the theory of cerebral localizations is what interests us least here; it is the least tenable aspect of Auguste Comte's thinking. We may omit this aspect because Comte himself stated that the cerebral localizations are, to a certain extent, hypothetical. Plato, too, had a theory of localizations, if not of the brain, at least of the body. After having distinguished the νοῦς, the θύμος, Plato locates these different aspects of human nature in the different parts of the body. But we can disregard this idea of the location of affections in the body if we wish to study only Plato's image of man.[25]

Comte indicated that human nature may be regarded as either twofold or threefold. If we regard man as twofold, he may be said to consist of a heart and a mind. If threefold, the heart will be divided into sentiment (or affection) and action (or will), and we will regard man simultaneously as sentiment, will, and intelligence. Comte adds that the double significance of the word "heart" is a well-founded ambiguity: to have heart is to have feelings or to have courage. The two ideas are expressed by the same phrase, as if language were aware of the existing bond between affection and courage (i.e., will).

Man is emotional, active, and intelligent; the problem is to understand the relations among these three elements. Comte's solution is as follows. Man is made for action; he is an active creature; and, returning at the end of his life to the idea that was already present in the *Opuscules,* Comte wrote frequently in the *Système de Politique Positive* that man is not made to waste his time in endless doubts and speculations. Man is made to act. But, made to act, he will never act through intelligence. Abstract thought will never be the deciding factor. The impulse to act will always come from the heart (in the sense of emotions). But this emotionally inspired activity requires the control of the intelligence. Or, to use one of Comte's famous formulas, man is *to act by emotion and to think in order to act.*

Out of this conception arises a critique of a certain kind of intellectualism or rationalism, according to which the historical process would make intelligence the deciding factor in human behavior. According to Comte, this can never be the case. It will always be emotion which is the source of the impulse. Emotion will always be, so to speak, the soul of humanity and the soul of action. The intelligence will never be more than an organ to guide, direct, and control.

But this is not to underrate intelligence, for in Comte's philosophy there is the notion of an inverse relationship within the human faculties, between strength and nobility. What is noblest is also weakest, and the fact that intelligence does not determine action does not belittle intelligence. Quite the contrary; intelligence is not and cannot be strength precisely because, in a sense, it is the noblest thing there is.

The cerebral localizations of these three elements of human nature are merely the transposition of ideas about their functioning. Comte placed the intelligence near the forward part of the brain, so that it is in direct communication with the organs of perception, the sense organs. Conversely, he placed the emotions behind, so that they may be in direct communication with the motor organs.

The breakdown of these three elements is as follows. Among the emotions, we can distinguish those which pertain to egoism and those which, on the contrary, pertain to altruism or disinterestedness. We arrive at a strange catalogue: First, the purely egoistical instincts or dispositions—the nutritive, sexual, and maternal instincts. Then Comte adds to the list those tendencies which are still egoistical, but which already pertain to relations with others—the military and industrial instincts. (This is the transposition into human nature of the two types of society he thought he had observed in his own age. The military instinct is the one that encourages us to overcome obstacles; the industrial instinct, on the other hand, is the one which induces us to produce goods.) And then Comte adds two words which we have no trouble recognizing: *pride* and *vanity*. Pride is the instinct to dominate; vanity, the search for the approval of others. So that with vanity we have already passed in a sense from egoism to altruism.

The nonegoistical tendencies are three in number: friendship—the feeling of one person for another, on an equal footing; veneration—which already widens the circle and which applies to the feeling of the son for the father, the disciple for the master, the inferior for the superior; and, finally, kindness—which in principle has universal application and which is to be fulfilled in the religion of humanity.

The breakdown of intelligence is as follows. The first distinction is between understanding and expression. Understanding, in turn, may be passive or active. When passive, it is abstract or concrete; when active, it is inductive or deductive.

Finally, the breakdown of the will is threefold: it consists of courage (to undertake), prudence (to execute), and steadfastness (to complete).

Such, briefly, is Comte's *tableau cérébral,* or his theory of human nature. History does not change this human nature; the primacy of statics is tantamount to a declaration of the timeless nature of those dispositions characteristic of man as man. Comte would not have agreed with Sartre that "man is his own future." He would not have said that man created himself through the ages; the essential dispositions are present from the beginning. But this does not mean that the succession of societies does not make its own contribution. This contribution is man's possibility of realizing what is noblest in his own nature, of promoting the gradual fulfillment of the altruistic tendencies. It is also man's possibility for utilizing his intelligence as a guide. Man's intelligence will never be anything but a means of control; but it could not be a dependable guide to action in the early days of history because, as we have seen, positivist thinking is not spontaneous thinking. Man cannot be instinctively a positivist. To be a positivist is to discover the laws governing phenomena, and it requires time to glean such understanding from observation and experimentation. Hence, history is necessary so that human intelligence may attain its intrinsic end, may realize its peculiar vocation.

The structural relationships between the elements of human nature will always remain what they have been from the beginning. I stress this point because Comte objected to a certain optimistic and rationalistic interpretation of human development. Contrary to those who imagined that reason could be the deciding factor in human behavior, Comte asserted that men will never be prompted by anything but their feelings. The goal is for men to be prompted more and more by unselfish feelings, and less and less by egoistic instincts, and for the organ of control that guides human behavior to fulfill its function by discovering the laws governing reality.

From this interpretation of human nature, let us now turn to the interpretation of social nature. The statics of the *Systéme de Politique Positive,* Part II, is presented in the following manner. Auguste Comte outlined, one after the other, a theory of religion, a theory of property, a theory of the family, a theory of language, a theory of the social organism or the division of labor; and he concluded with two chapters, one devoted to the social order systematized by the priesthood (a rough draft of human society become positivist), and the other devoted to the limits of variations —a static interpretation of the possibility of historical dynamics. These chapters form a theory of the fundamental structure of societies.

The chapter devoted to religion is an attempt to demonstrate the function of religion in every human society. Every society necessarily involves consensus, agreement between groups, agreement between individuals, a unifying principle. Religion is this unifying principle. Religion contains within itself the threefold division characteristic of human nature. It includes an intellectual aspect, dogma; an affective aspect, love; and a practical aspect, which Comte calls the regime or the cult. Religion imitates the divisions of human nature because, in order to create unity, it must address itself simultaneously to the intelligence, the emotions, and the will.

This conception is not fundamentally different from the one Comte had expounded at the beginning of his career, when he maintained that the ideas of the intelligence determined the stages of the history of mankind. But at this point in his life he no longer saw simple ruling ideas or philosophy as the foundation of every social order, and religion is affectivity and activity as well as dogma or creed.

From the chapter on religion, let us proceed to the two chapters pertaining to property and language, respectively. The juxtaposition may seem strange, but it corresponds to Comte's thinking.[26] There is an analogy between property and language. Property might be called the projection of activity onto society, while language is the projection of the intelligence. But the law common to property and language is what might be called the law of accumulation. Civilization exists because material and intellectual conquests do not vanish with the conquerors. Man exists only by virtue of tradition, i.e., the transmission of goods and knowledge. Property is the accumulation of goods transmitted from one generation to another. And language is, so to speak, the receptacle in which the acquisitions of the intelligence are preserved. When we inherit a language, we are inheriting a culture created by our ancestors.

Do not be misled by the word *property*, with all its political or partisan overtones. In Comte's eyes, it is of no importance whether property be private or public. For him, property, inasmuch as it is an essential function of civilization, represents the fact that man's material productions outlive their creators. We transmit to our descendants what we have produced, what we have created. The two chapters on property and language are devoted to the two essential agents of human civilization, which requires the continuity of generations and the adoption by the living of the thought of the dead. Whence the familiar statement: "Human society has more dead than living members."

These formulas deserve some attention. One of the things that make Comte so original is that, starting from the idea of an industrial society, convinced that scientific societies differ fundamentally from the societies of the past, he arrived (unlike the majority of modern sociologists), not at the depreciation of the past and the exaltation of the future, but at a kind of rehabilitation of the past. A utopian, dreaming of a future more perfect than any known society, he remains a man of tradition with a keen sense of human unity throughout the ages.[27]

Between the chapter on property and the one on language, Comte inserted a chapter on the family, which parallels the one on the social organism or the division of labor. These two chapters correspond to two of the elements of human nature. The family is essentially the affective element, while the social organism or division of labor corresponds to the active element in human nature.

Comte constructed a theory of the family in which he takes as his model—and implicitly regards as exemplary—the Western type of family, and naturally he has been criticized for this. He arbitrarily discarded as pathological certain family systems which have existed down through the ages—for example, polygamy. There is no doubt that Comte was systematically arbitrary and that in his treatment of the family he frequently confused certain traits belonging to a particular society with universal traits. But I do not believe this easy criticism exhausts the matter. Comte tried hard to show that the various relations existing within the family correspond to the various relations which may exist between human beings, and also that in the family, human affectivity was trained and molded.

Here is the scheme of family relations: a relation of equality between brothers; a relation of veneration between children and parents; a complex relation of authority—obedience between man and wife. According to Comte, the husband obviously has the authority, but it is an authority which is to a certain extent inferior, because it is the authority of man, or activity and intelligence, over woman, who is essentially sensibility; this supremacy, based in a sense on strength, is from another point of view inferiority, because in the family the spiritual power—that is, the noblest power—belongs to woman.

Auguste Comte had, if you will, a sense of the equality of beings, but it was an equality based on the radical differentiation of functions and natures. When he said that woman is intellectually inferior to man, he was ready to see this as a superiority of woman, because by the same token woman is the spiritual power, the power of love, which to the Comte of the *Système* was far more important than the futile superiority of intelligence. At the same time, in the family it is the men who have the experience of historical continuity, who learn what is the condition of civilization, who control the transmission of civilization from generation to generation.

As regards the division of labor, Comte's essential idea is that of the differentiation of activities and the co-operation of man. The exact terms are: separation of functions and combination of efforts. But his profoundest idea, and probably one of Comte's most original in this respect, is the recognition of the primacy of force in the practical organization of society. Society, conceived as a way of organizing activity, is dominated, and cannot help being dominated, by force.

Auguste Comte acknowledged only two political philosophers, Aristotle and Hobbes. The only political philosopher between Aristotle and himself who, in Comte's opinion, deserved (or almost deserved) to be mentioned is Hobbes. Why? Because Hobbes saw clearly that every society is ruled, must be ruled (*must* in both senses of *inevitable* and *desirable*), by force. What is force in society? Number, or wealth.[28] Thus no utopias, no illusions, no idealism: society is and will be dominated by the forces of number or of wealth (or a combination of these—there is no fundamental qualitative difference between the two). That force should prevail is normal. How could it be otherwise as long as we look at life as it is, at human society as it is?

But a society consistent with human nature must include a complement to the domination of force, just as in human nature there must be a complement to the inevitable primacy of the affective impulses. There is, then, as a counterpart to the realistic theory of the social order which recognizes the domination of force, a Comtist theory of spiritual power. Spiritual power is a constant necessity in human societies because the latter, as a temporal order, will always dominate by force.

This spiritual power is twofold: that of the intelligence and that of the feelings or the affections. At the beginning of his career, Comte interpreted spiritual power as primarily that of the intelligence. At the end of his career, spiritual power was primarily that of the affections or of love. But, whatever the precise form assumed by spiritual power, the distinction between temporal and spiritual power is of all times, of all ages, though it is not fully realized until the positive phase, the culmination of human history.

Spiritual power has several functions. It must regulate the inner life of man. It must rally men to live and act in common. It must sanctify the temporal power in order to convince men of the need for obedience, because social life is impossible unless there are men who command and others who obey. (After all, it is of little importance, at least in the eyes of philosophers, to know who commands and who obeys, because in the last analysis those who command are and always will be the powerful of this world.) But spiritual power must not only regulate, rally, and sanctify; it must also mitigate and limit temporal power. In order for it to do this, social differentiation must be already at an advanced stage. When the spiritual power sanctifies the temporal power—that is, when the priests declare that the kings are God's anointed or that they rule in God's name—the spiritual power adds to the authority of the temporal power. This sanctioning of the strong by the spirit may have been necessary in the course of human history. Naturally, there had to be a social order, and an accepted social order, even before the mind had discovered the true laws of the physical order, let alone the true laws of the social order. But, in the final phase, the spiritual power will bestow a partial consecration on the temporal power: the scientists will justify the industrial order, and in so doing they will add a kind of moral authority to the ruling power of management, or the bankers. But their essential function will be not so much to sanctify as to temper and limit, that is, to remind the powerful that they are merely performing a social function and that, further, their leadership implies no moral or spiritual superiority.

In other words, in the static analysis of the distinction between temporal and spiritual power, we discover that history is necessary so that the spiritual power may fulfill all its functions and so that the true distinction between temporal and spiritual may be finally recognized and applied.

Thus, we may conclude this analysis of statics by clarifying the meaning of dynamics from the threefold standpoint of intelligence, activity, and sentiment. The history of intelligence proceeds from fetishism to positivism, i.e., from the synthesis based on subjectivity, and on the projection onto the external world of a reality comparable to that of consciousness, to the discovery and establishment of the laws governing phenomena without claiming to reveal their causes. Activity proceeds from the military phase to the industrial phase—that is, in Marxist terminology, from the struggle of men among themselves to the victorious struggle of man with nature (with this reservation, however, that Comte nourished no exaggerated hopes about the results of man's mastery over natural forces). Finally, as regards affectivity, history consists of the gradual fulfillment of altruistic tendencies, without man's ever ceasing to be instinctively and primarily egoistic.

History, therefore, leads simultaneously to an increasing differentiation of social functions and an increasing unification of societies. In the final phase, temporal and spiritual will be more distinct than they have ever been, and this distinction will at the same time be the condition of a more profound consensus, of a deeper unity. Men will accept the temporal hierarchy because they will recognize its precariousness and will reserve their highest appreciation for the spiritual order which both sanctions and limits the temporal hierarchy.[29]

V

IN THE LAST three sections I have discussed Comte's three versions of the central idea of all his writings; that is, the unity of the human race. First, we stressed the typical features of industrial society as the form of social organization capable of becoming universal. Next, in the *Cours de Philosophie Positive,* we saw that the history of mankind may and must be considered as the history of a single people. And finally, because the history of mankind is that of a single people, it follows that this unity of species is based on the constancy of human nature and finds expression in a fundamental order which may be recognized in spite of the diversity of institutions which history presents.

At the same time, the sociologist of human unity must have a philosophical attitude underlying his sociology. To conclude this brief survey of Comte's thought, I think it would be useful to complete the picture of Comte the philosopher, the outlines of which I have already sketched.

Auguste Comte is a sociologist among philosophers and a philosopher among sociologists. The indissoluble link between sociology and philosophy has its source in Comte's first principle, namely, the affirmation of human unity, which implies a certain conception of man, his nature, his destiny, and the relation between individual and collectivity. In order to try to reveal, not all, but several of Comte's philosophical ideas, it seems to me a good idea to relate his thinking as I have summarized it to the three aims which may be found in his work: the aim of the social reformer, the aim of the philosopher synthesizing the methods and results of the sciences, and, finally, the aim of the man who appoints himself high priest of a new religion, the religion of humanity—in which man as the Sublime Being is the object of love and worship.

One way or another, most sociologists have been concerned with action. All the major sociological doctrines of the nineteenth century imply a transition from thought to action, or from science to politics and ethics. This raises such questions as: How does the sociologist effect the transition from theory to practice? What sort of practical advice may be derived from sociology? Is there an over-all solution to the social problem as a whole, or only partial solutions to a number of specific problems? Finally, once a solution is arrived at, how does the sociologist imagine that he will transform it into a reality?

If we had raised these questions with respect to Montesquieu, it seems to me that the answer would be this: Montesquieu sought to understand the diversity of social and historical institutions. He was very cautious about making the transition from science, which is concerned with understanding, to politics, which is disposed to direct or advise. To be sure, there are occasional suggestions to legislators, but Montesquieu's own preferences regarding various aspects of social organization have been, and still are, subject to debate. In any case, even if Montesquieu did give advice, he was a counselor who condemned certain practices rather than advised what should be done; the lessons which Montesquieu drew from sociology are negative rather than positive. He would, for instance, suggest that slavery is contrary to human nature, that some sort of equality between men is part of the very essence of humanity. But, as soon as he treated a particular society at a certain period, the supreme counsel to be drawn from his works will be: examine these people, observe the milieu in which they live, consider their history, do not overlook their character, and try to use common sense. A fine program, but not one of any great precision; the looseness is, moreover, typical of

Montesquieu, who did not dream of an over-all solution to what in the nineteenth century was referred to as "the crisis of our civilization" or "the social problem."

In other words, Montesquieu conceived only the most cautious and limited translation from science to action. He suggested partial solutions, and not one over-all solution. He did not advocate the use of violence in order to make existing societies conform to his idea of the just order. He had no miraculous recipe whereby the prince is prudent and wise and the prince's counselors have read *The Spirit of the Laws*. Montesquieu was modest.

Modesty was certainly not the outstanding quality of Auguste Comte as a social reformer. Unlike Montesquieu, he possessed the solution to the problem of society. Since human history is one and essential order underlies its variations, he was able to conceive beforehand what the fulfillment of human destiny, and the perfect realization of the social order, must be.

In this conception, Comte neglected economics and politics in favor of science and ethics. The scientific organization of labor is necessary. He felt toward most reforming ideas the twofold scorn of the man of science and the founder of a religion. He was convinced that societies have the governments they deserve, those which correspond to the state of their social organization. He did not believe that by changing the government and the constitution man puts an end to the underlying troubles of society. He was a social reformer; he did want to eliminate the traces of the feudal and theological mentality, to convince his contemporaries that war is obsolete and colonial conquest absurd. But to him these truths were so obvious that he did not concentrate on demonstrating them. His first concern was to spread a way of thinking which would automatically lead to the just organization of society and of the state. To make all men positivists, to make them understand that the positivist system is the logical solution to the problem of the temporal order, to inculcate disinterestedness and love with regard to the spiritual or moral order—that is his goal.

The paradox is that this essential order which Auguste Comte sought to achieve must, according to his philosophy, be achieved of itself. For if the laws of positivist statics amount to an unchanging order, the laws of positivist dynamics give us assurance that the essential order will fulfill itself. This is a contradiction which occurred in another form in Marx's thinking, but Comte's solution was quite different.

Like Montesquieu—even more than Montesquieu—Auguste Comte was opposed to violence. He did not believe that the solution to the crisis of modern society lies in revolution or civil war, or that by these means societies will fulfill their destinies. He admitted that it will take time to effect the transition from today's lacerated societies to the reconciled societies of tomorrow. But at the same time he justified action, and the efforts of men of good will, by means of what he called the modifiable character of fatality. History is subject to laws, and thanks to Auguste Comte, we are not ignorant of the order toward which human societies are spontaneously evolving. But this evolution may take more or less time and cost more or less bloodshed. In the length and modes of that evolution, itself inevitable, is expressed man's share of freedom. According to Comte, the higher one rises on the ladder of being, from the simplest to the most complex, the wider the margin of freedom—or, to use his barbarous expression, the margin of modifiability of fatality. Now, what is most complex is society—or still more, the individual social being, object of the seventh science, i.e., ethics, in his latest classification of the sciences. Hence it is in human history that the laws allow men the greatest freedom.[30]

The sociologist-social reformer, according to Comte, is not, therefore, the engineer of particular reforms. Nor is he the prophet of violence, like Marx. Comte was the serene herald of the new order. The sociologist is a sort of peace-loving prophet who trains our minds, rallies our souls, and secondarily is himself the high priest of the sociological (or sociographical) religion.

Let us turn now to our second theme, namely, sociology and the synthesis of the sciences. As we have seen, from his youth Comte had two ruling ideas: one was to reform society, and the other to establish the synthesis of all scientific knowledge. The connection between these two ideas should be perfectly clear by now. The only social reform worth the effort is one which would transcend the theological and feudal way of thinking and propagate the positivist attitude. But this reform of collective thinking can only be the result of scientific progress. And the best way to establish the new science is to trace the progress of the positive spirit throughout history and into the existing sciences.

Comte's synthesis of the sciences is possible only because of a conception of science which, in turn, is intimately related to his aims as a reformer and a sociologist. I should like to indicate a few of these Comtist views of science which explain the transition from the positivism of his first period to the positivism of the last, from the *Cours* to the *Système,* and which help us understand why a number of positivists who had followed Comte during the first part of his career felt that at the end of his life Comte had contradicted himself.

Science, as Comte conceives it, is not an adventure. It is not an endless and infinite quest. Science is a source of dogmas. Comte wanted to eliminate the last traces of theological spirit, but in a sense he was born with certain theological pretensions. He was looking for truths acquired once and for all, never again to be brought in question. One of his convictions was that man is made, not to doubt, but to believe. The sciences lead us to sociology largely because they provide us with a body of truths, acquired once and for all, which are equivalent to the dogmas of the past.

Second, Comte believed that the essence of scientific truth is represented by what he called laws, that is, according to his system, either inevitable relations among phenomena or dominant and constant facts characteristic of a certain kind of being.

Comte's science is not a quest for a final explanation; it makes no claim to search out causes. It is content to observe the order that prevails in the world, not so much through disinterested curiosity as to take advantage of the resources nature offers and to establish order in our own minds. Hence science, as Auguste Comte conceived it, is pragmatic in two senses. It is pragmatic because it is the source from which technological solutions are drawn as inevitable consequences; it is pragmatic because it has an educational value in relation to our intelligence, or rather in relation to our consciousness. Our consciousness itself would be chaotic, our subjective impressions—to use Comte's terminology—would mingle in a confused way and would produce nothing intelligible, were there not external to us an order which we discover and which is the origin and source of the order of our intelligence.[31]

I have no doubt that the founder of positivism would be outraged by the sputniks, by the presumption of exploring space beyond the solar system. He would consider such an undertaking senseless: Why go so far when we do not know what to do where we are? Any science which did not have the virtue of revealing an order or enabling us to act was in his eyes useless and therefore undesirable.

This conception of science logically brings us to sociology and ethics, as the culmination and fruition of the intrinsic purpose of science. If science were a search for the truth, an endless quest for explanation, a desire to grasp a meaning which escapes us, it might more closely resemble what science is in reality, but it would not be so likely to lead to sociology as the dogmatic-pragmatic science conceived by Auguste Comte.

Third, when Comte tried to unite the results and methods of the sciences, he discovered (or thought he discovered) a structure of reality which is necessary to man's understanding of himself and the sociologists' understanding of societies. I refer to the hierarchical structure of beings, in which each kind of creature is subject to laws. The governing idea of this hierarchical view of the world is that the inferior conditions, but does not determine, the superior. There is a hierarchy in nature, from the simplest to the most complex phenomena, from inorganic nature to organic nature, and, finally, to living creatures and man. Basically, this structure, even if it demands a certain development in order to be realized, is a quasi-immutable one. It is the hierarchy laid down by nature. This hierarchical view enables us to locate social phenomena in their proper places and at the same time to determine the social hierarchy itself.

Fourth, the contemporary sciences, which are the expression and fulfillment of the positive spirit, which must provide the dogmas of modern society, were not therefore free from an ever-present danger inherent in their nature: the danger of dispersion by analysis. Comte unceasingly reproached his scientific colleagues for a double specialization which seemed to him to be excessive. The scientists studied one little section of reality, one little area of a science, and ignored the rest. This is scientific or analytical specialization, so to speak. Further, the scientists were not all as sure as Comte was that they represented the priests of modern society and that they ought to have exercised a spiritual authority. They were lamentably inclined to be content with their role as scientists, without ambition to reform the world. Deplorable modesty, said Comte; a fatal aberration. The purely analytical sciences would be more harmful than useful in the end. What is the point of an endless accumulation of facts? There must be a synthesis. It is Auguste Comte, naturally, who will work out the synthesis of the sciences. But it must be understood that this synthesis of the sciences has its center or origin in sociology itself. Indeed, one might say that all the sciences converge on sociology, because the whole hierarchy of being culminates in the human species, which represents the highest level of complexity, nobility, and fragility among beings.

Thus, when Comte established the synthesis of the sciences, to culminate in sociology, he was merely following, as it were, the natural direction of the sciences, which tended toward the science of society as toward their end, in the double sense of conclusion and goal.

Not only does the synthesis of the sciences function objectively in relation to sociology, the science of the human species, but the only possible subjective principle for synthesis is, again, sociology. Why? Because the assembling of knowledge and methods is possible only with reference to human beings. If one was inspired purely and simply by curiosity, one might be content to observe the diversity of phenomena and relationships *ad infinitum*. For there to be a synthesis, one must consider objectively the hierarchy of being ascending to the human species, and one must consider subjectively the knowledge that has a relation to man, whose condition is explained—the knowledge which is useful to him both in exploiting natural resources and in living according to the just order.

Thus there is, after all, a sort of primary philosophy, as Comte called it, in the *Système de Politique Positive*, with fifteen laws—some objective, others subjective—which help to explain how the sociologist synthesizes the findings of the sciences, because the sciences can achieve unity only in relation to mankind, objectively as well as subjectively.[32]

This enables us to propose another formulation: sociology, for Comte, is the science of the human mind. Man understands the human mind only on condition that he observe its activity and its productions throughout history and in society. One does not come to know the human mind either through introspection, in the manner of the psychologists, or by the method of reflexive analysis, in the manner of Kant.

The true science of the human mind is what we would today call the sociology of knowledge. The true science of the human mind is the observation, analysis, and comprehension of the capacities of the human mind as they are revealed to us through their productions in the course of history.

Sociology is also the science of the human mind, because the mind's way of thinking and activity are at every moment inseparable from the social context. There is no transcendental, timeless self which can be grasped by means of reflexive analysis. The mind is social; it is historical: the mind of each age, the mind of each thinker, is caught in a social context.

But, in addition to achieving a synthesis of the sciences, Auguste Comte saw himself as the founder of the religion of humanity. He believed that the religion of our age may and must be of a positivist inspiration. It can no longer be the religion of the past, because the latter presupposes a way of thinking that is outmoded. The man of scientific mind, said Comte, can no longer believe in revelation, in the catechism of the Church, or in divinity according to the traditional conception. On the other hand, religion answers a permanent need in man. Man needs religion because he needs to love something greater than himself. Societies need religion because they need spiritual power, which at once sanctifies and shapes the temporal power. Only a religion can restore to its place the technical hierarchy of ability and superimpose on it a (possibly opposing) hierarchy of merit.

What, then, will this religion be which will answer these perpetual needs of a humanity in search of unity and love? It will be the religion of humanity itself. But let us consider this. The hierarchy of moral worth which must be created may be opposed to the temporal hierarchy. The humanity Comte asked us to love is not humanity as it is, in all its injustice and vulgarity. The humanity he asked us to love is not all men, but those men who survive in their descendants, those who have lived in such a way as to leave works or examples.

If Comte's "humanity" consists of more dead than living members, it is not because there are statistically more dead than living men; it is because only the dead survive as the humanity we must love, the humanity worthy of what Comte called subjective immortality.[33]

In other words, the Great Being, the humanity Comte asked us to love, is the best men have had or done; in the end it is, in a sense, what transcends men in man.

Is this essential humanity which we love in the form of the Great Being so different from the humanity achieved and transcended in the God of the traditional religions? I realize that there is a fundamental difference between loving humanity, as Comte asked us to do, and loving the transcendent God of the traditional religions. But the God of Christianity became man; and the relation between essential humanity and divinity in the Western religious tradition lends itself to various interpretations. For my part, I believe Comte's religion—which, as you know, has not had a great worldly success—is not so absurd as is generally believed. In any case, it seems to me superior in the main to many other religious or semireligious ideas which other sociologists have circulated, either voluntarily or involuntarily. Surely it is better to love the essential humanity, of which great men are the expression and symbol, than to cherish an economic and social order to the point of mortally detesting all those who do not believe in its sanctity. If one insists on deriving a religion from sociology (which I do not), the only one that seems to me thinkable, were I forced to do so, is that of Auguste Comte, because it does not instruct us to love one society among others, which would be tribal fanaticism, or to love the social order of the future, which no one knows and in whose name one begins by exterminating all skeptics. What Comte wanted us to love is neither the French society of today, nor the Russian society of tomorrow, nor the American society of the day after tomorrow, but the essential humanity which certain men have been able to achieve and toward which all men should raise themselves. Perhaps this is not a "love object" which readily affects most men; but of all the sociological religions, Comte's sociocracy seems to me philosophically the best.

Of course, this may be the reason why it has been politically the weakest. It is difficult for men to love what would unite them and not to love what divides them, once they no longer love transcendent realities.

We know that Auguste Comte would probably not have conceived the religion of humanity if he had not been in love with Clotilde de Vaux. We are, therefore, free to regard this religion as a biographical accident. But this biographical accident does not, nevertheless, seem to me to lack significance, if my basic interpretation of Comte's thinking is correct. For, as I have said, Comte is the sociologist of human unity and one of the possible, if not inevitable, results of this sociology of human unity is the religion of human unity. Comte wanted men, though they are destined to live indefinitely in separate temporal societies, to be united by common convictions and by a single object of their love. Since this object could have no transcendent existence, was there any other solution but to imagine men united in the worship of their own unity by the desire to achieve and to love that which, regardless of centuries and cultures, transcends all particularity?

BIOGRAPHICAL CHRONOLOGY

1798   January 19. Birth of Auguste Comte in Montpellier of a Catholic and Monarchist family. His father was a senior official, proxy for the Chief Collector's Office in Montpellier.

1807–14   Attended the *lycée* in Montpellier. Comte soon broke away from the Catholic faith and adopted liberal and revolutionary ideas.

1814–16   Studied at the Ecole Polytechnique in Paris, to which he was the first of the southern applicants to be admitted.

1816   In April the Restoration Government decided to close provisionally the Ecole Polytechnique, which was suspect of Jacobinism. Returning to Montpellier for several months, Comte took courses in medicine and physiology from the city's Faculty of Medicine. He then went back to Paris where he earned a living by teaching mathematics.

1817   In August, Comte became Saint-Simon's secretary,

remaining his colleague and friend until 1824. During this time he was also associated with various publications on the philosophy of industrialism: *L'Industrie, Le Politique, L'Organisateur, Du Système industriel, Catéchisme des industriels.*

1819   *Séparation générale entre les opinions et les désirs,* an article that appeared in *Censeur* edited by Charles Comte and Charles Dunoyer.

1820   *Sommaire appréciation sur l'ensemble du passé moderne,* published in the April issue of *L'Organisateur.*

1822   *Prospectus des travaux scientifiques nécessaires pour réorganiser la Societé,* published in *Système industriel.*

1824   *Système de politique positive,* Vol. I, Part I, revised edition of the preceding work.

In April Comte sold this piece to Saint-Simon who then had it published anonymously in *Catéchisme des industriels.* Comte protested, and they had a falling-out. H. Gouhier wrote: "Comte's employer looked on the piece as the third part of *Catéchisme des industriels,* which dealt with the industrialism of Saint-Simon. Comte himself saw it as the first part of his *Système de politique positive,* which set forth his ideas on positivism." Comte was to speak from then on of "the disastrous influence" worked on him by "a fatal relationship" with a "depraved charlatan."

1825   *Considérations philosophiques sur les sciences et les savants, Considérations sur le pouvoir spirituel.* These two pieces were published in spite of what had passed in Saint-Simon's *Le Producteur.*

Comte married Caroline Massin, a former prostitute. The marriage—the result of "charitable judgment"—was, said Comte, "the only serious mistake of my life." Caroline Massin was to leave him several times.

1826   In April he began teaching a course in Positive Philosophy. Humbolt, H. Carnot, the physiologist Blainville, and the mathematician Poinsot were among his students.

1826–27   After his wife left him for the first time, Comte suffered a severe breakdown and was sent to a sanitarium. After eight months he left, not yet cured, and attempted suicide shortly thereafter. The crisis passed. Comte, well aware of its cause, imposed a

	rigorous physical and mental routine on himself to try to ward off a recurrence.
1829	Comte took up his course in Positive Philosophy again on January 4.
1830	Publication of Volume I of *Cours de philosophie positive.* The other volumes appeared in succession in 1835, 1838, 1839, 1841, and 1842.
1831	Began a free course in popular astronomy, which he taught in the Town Hall of the Third Arrondissement and which he continued to teach until 1847–48. Comte unsuccessfully sought the lectureship in Analysis at the Ecole Polytechnique.
1832	Appointed Assistant Lecturer in Analysis and Mechanics at the Ecole Polytechnique.
1833	Comte wanted Guizot to create a lectureship in the History of Science for him at the Collège de France. He was refused.
1836	Appointed Admissions Examiner at the Ecole Polytechnique.
1842	Final separation from Madame Comte.
1843	*Traité élémentaire de géometrie analytique.*
1844	*Discours sur l'esprit positif,* preamble to *Traité philosophique d'astronomie populaire.*
	Comte lost his position as Examiner at the Ecole Polytechnique. He then lived off a "free Positivist subsidy" sent him first by John Stuart Mill and some rich Englishmen in 1845 and then by E. Littré and a hundred or so followers or French admirers from 1848 on.
	In October Comte met thirty-year-old Clotilde de Vaux, sister of one of his former students, who was separated from her husband and aware that she was fatally ill.
1845	"The year without equal." Comte confessed his love to Clotilde de Vaux, who offered only friendship, declaring herself "powerless for anything beyond the limits of affection."
1846	April 5. Clotilde de Vaux died under Comte's very eyes. From that moment on he dedicated himself to religion.
1847	Comte proclaimed Humanism.
1848	Founding of the Positivist Society.
	Discours sur l'ensemble du positivisme.
1851	Comte lost his position as Assistant Lecturer at the Ecole Polytechnique.
	Publication of the first volume of *Système de poli-*

*tique positive ou Traité de sociologie instituant la
religion de l'humanité.* The other volumes appeared
in 1852, 1853, and 1854.

Comte wrote Monsieur de Thoulouze on April 22:
"I am convinced that I will preach positivism as the
only real and complete religion at Notre-Dame before 1860."

In December, Littré and several other admirers,
shocked by Comte's approval of Louis-Napoleon's
coup d'état and apprehensive about the new philosophical trend, resigned from the Positivist Society.

1852 *Catéchisme positiviste ou sommaire exposition de la
religion universelle.*

1855 *Appel aux conservateurs.*

1856 *Synthèse subjective ou système universel des conceptions propres a l'état normal de l'humanité.*

Comte proposed to the Jesuit Superior an alliance
against "the anarchic eruption of Western delirium."

1857 September 5. Comte died in Paris, at 10, rue
Monsieur-le-Prince, surrounded by his followers.

NOTES

1. Auguste Comte conceived the law of the three stages in
February or March of 1822 and expounded it for the first time
in the *Prospectus des travaux scientifiques nécessaires pour réorganiser la Société,* published in April 1822 in a volume by
Saint-Simon entitled *Suite des travaux ayant pour objet de
fonder le système industriel.* This work, which Comte was to
call, in his preface to the *Système de politique positive,* the
Opuscule fondamental and which is sometimes referred to as
the *Premier système de politique positive,* after the title of the
1824 edition, reappeared in Volume IV of the *Système de politique positive* under the title *Plan des travaux scientifiques nécessaires pour réorganiser la Société.*

The law of the three stages is the subject of the first lecture of the *Cours de philosophie positive* (Fifth Edition, Vol. I,
pp. 2–8), and the classification of the sciences is the subject of
the second lecture in the same *Cours* (*Ibid.,* pp. 32–63).

On the discovery of the law of the three stages and of the
classification of the sciences, see especially Henri Gouhier, *La
Jeunesse d'Auguste Comte et la formation du positivisme,* Vol.
3, *Auguste Comte et Saint-Simon,* Paris, Vrin, 1941, pp. 289–91.

2. "In studying the total development of the human intelligence in its various spheres of activity from its first and simplest

flights down to our own time, I believe I have discovered a great and fundamental law to which this development is bound by an invariable necessity and which seems capable of being solidly established, either upon the rational proofs furnished by an understanding of our organization, or upon the historical verifications resulting from an attentive examination of the past. This law consists in the fact that each of our principal conceptions, each branch of our understanding, passes successively through three different theoretical stages: the theological or fictitious state; the metaphysical or abstract state; and the scientific or positive state. In other words, the human mind, by its very nature, employs successively in each of its fields of investigation three methods of philosophizing whose character is essentially different and even radically opposed: first the theological method, next the metaphysical method, and finally the positive method. This gives rise to three kinds of philosophy, or of general conceptual systems about all phenomena which are mutually exclusive. The first is the necessary point of departure of the human intelligence; the third is its fixed and definitive stage; the second is destined to serve solely as a transition. . . .

"In the positive stage, the human mind, recognizing the impossibility of arriving at absolute notions, renounces the quest for the origin and destiny of the universe and the attempt to know the underlying causes of phenomena, and devotes itself to discovering, by means of a judicious combination of reason and observation, their actual laws, that is, their invariable relations of succession and similitude. The explanation of facts, thus reduced to their real terms, is henceforth nothing but the relation established between the various particular phenomena and a few general truths whose number the advances of science tends increasingly to diminish." (*Cours de philosophie positive*, Vol. I, pp. 2–3.)

3. Comte writes: "Ideas govern the world or throw it into confusion; in other words, the whole social mechanism rests ultimately on opinion. . . . The great political and moral crisis of present societies stems, in the last analysis, from intellectual anarchy. Our gravest malady consists, in effect, in this profound divergence which now exists among all minds with respect to all those fundamental maxims whose permanence is the first condition of a true social order. So long as individual intelligences have not accepted, by unanimous consent, a certain number of general ideas capable of forming a common social doctrine, there is no escaping the fact that the state of nations will necessarily remain essentially revolutionary, despite all the political palliatives that may be adopted, and will actually be

characterized only by provisional institutions." (*Cours de philosophie positive,* Vol. I, p. 26.)

4. However, Comte does not overlook its importance. "Industrial life gives rise only to classes which are imperfectly associated among themselves, for want of an impulsion sufficiently general to coordinate everything while disturbing nothing; which constitutes the principal problem of modern civilization. The true solution will become possible only if it is based on civic cohesion." (*Système de politique positive,* Vol. III, p. 364.) "Since the abolition of personal servitude, the proletarian masses are not yet, apart from all anarchical oratory, truly incorporated into the social system; . . . the power of capital, at first a natural means of emancipation and next of independence, has now become exorbitant in daily transactions whatever just preponderance it must necessarily enjoy by virtue of a superior generality and responsibility, according to sound hierarchical theory." (*Cours de philosophie positive,* Vol. VI, p. 512.) "The principal disorder today affects material existence, in which the two necessary elements of the ruling force, number and wealth, live in a growing state of mutual hostility, for which they must each be reproached." (*Système de politique positive,* Vol. II, p. 391.)

5. James Burnham, *The Managerial Revolution,* New York, 1941.

6. Comte's examination of the nature and purpose of political economics is found in the forty-seventh lecture of the *Cours de philosophie positive* (Vol. IV, pp. 138 ff.). Auguste Comte had known and studied the political economics of his time, that is, classical and liberal economics, when he was secretary to Saint-Simon, and in his critical writings he sets aside "the eminently exceptional case of the illustrious and judicious philosopher Adam Smith." His polemic is aimed primarily at Smith's successors: "If our economists are truly the scientific successors of Adam Smith, let them show us, then, in what way they have actually perfected and completed the doctrine of this immortal master, what truly new discoveries they have added to his brilliant original insights, which have, on the contrary, been essentially distorted by a useless and infantile display of scientific forms. When one views with an impartial eye the sterile disputes that divide them on the most elementary notions of *value, utility, production,* etc., one would think one was witnessing the most curious debates of medieval scholastics about the fundamental attributes of their pure metaphysical entities, which economic conceptions tend increasingly to resemble, the more dogmatic and subtle they become." (*Ibid.,* p. 141.) But the fundamental reproach that Comte addresses to the economists is

that they wanted to create an autonomous science, "isolated from the whole of social philosophy . . ." "For, by the nature of the subject, in the social studies, as in all studies pertaining to living bodies, the various general aspects are of necessity mutually interdependent and rationally inseparable, to the point where one aspect cannot properly be illuminated by another. . . . When one leaves the world of entities and turns to real speculations, it becomes clear that the economic or industrial analysis of society cannot be positively accomplished apart from its intellectual, moral and political analysis, either in the past or even in the present; so that, reciprocally, this irrational separation offers unimpeachable evidence of the essentially metaphysical nature of the doctrines resting on its foundation." (*Ibid.*, p. 142.)

7. Comte writes as follows: "After explaining the natural laws which, in the system of modern sociability, must determine the indispensable concentration of wealth among industrial leaders, positive philosophy will show that it matters little to popular interests in whose hands capital is habitually found, provided its normal use is necessarily useful to the social mass. But this essential condition depends by its nature much more on moral means than on political measures. In vain would narrow views and venomous passions legally establish elaborate impediments against the spontaneous accumulation of capital, at the risk of directly paralyzing all real social activity; it is clear that these tyrannical procedures would have much less real effectiveness than the universal reprobation directed by positive morality at any overly selfish use of the wealth possessed, a reprobation all the more irresistible in that the very persons who would have to suffer it could not dismiss its principle, which is inculcated in all by the common fundamental education, as Catholicism showed in the time of its preponderance. . . . But, by indicating to the people the essentially moral nature of their most serious demands, the same philosophy will necessarily communicate to the upper classes the weight of such a judgment by energetically imposing on them, in the name of principles which are no longer openly debatable, the great moral obligations inherent in their position; so that, on the subject of property, for example, the rich will regard themselves morally as the necessary depositaries of the public capital, whose actual use, without ever involving any political responsibility, except in a few exceptional cases of extreme aberration, must nevertheless always remain subject to a scrupulous moral debate which is necessarily accessible to all under the proper conditions and whose spiritual authority will later constitute the normal organ. By an extensive study of modern

evolution, positive philosophy will show that, since the abolition of personal servitude, the proletarian masses are not yet, apart from all anarchical oratory, truly incorporated into the social system; that the power of capital, at first a natural means of emancipation and next of independence, has now become exorbitant in daily transactions, whatever just preponderance it must necessarily enjoy by virtue of a superior generality and responsibility, according to sound hierarchical theory. In brief, this philosophy will show that industrial relations, instead of remaining subject to a dangerous empiricism or an oppressive antagonism, must be systematized according to the moral laws of a universal harmony." (*Cours de philosophie positive*, Vol. VI, pp. 357–58.)

8. Maurras wrote, among other things, an essay on Auguste Comte which was published along with some other essays (*Le Romantisme féminin, Mademoiselle Monk*) following *L'Avenir de l'intelligence*, Paris, Nouvelle Librairie Nationale, 1918. Maurras writes of Comte: "If it is true that there are masters, if it is false that heaven and earth and the means of interpreting them did not come into the world until the day of our birth, I know the name of no man that must be pronounced with a keener sense of gratitude. His image cannot be evoked without emotion. . . . Some among us were a living anarchy. To them he restored order, or the hope of order, which amounts to the same thing. He showed them the beautiful face of unity, smiling in a sky which does not seem too remote."

9. References to Comte are constant in the work of Alain. See especially: *Propos sur le christianisme*, Paris, Rider, 1924; *Idées*, Paris, Hartmann, 1932, reissued in the collection 10/18, Paris, Union Générale d'Editions, 1964 (this last volume contains a study devoted to Comte).

Alain's politics is set forth in the two volumes: *Eléments d'une doctrine radicale*, Paris, Gallimard, 1925; and *Le Citoyen contre les pouvoirs*, Paris, S. Kra, 1926.

10. I discussed the subject of war in the thought of Auguste Comte in *La Société industrielle et la guerre*, Paris, Plon, 1959, especially the first essay, which was the text of an Auguste Comte Memorial Lecture delivered at the London School of Economics and Political Science.

11. Several years ago I served on a doctorate committee for a thesis on Alain by a man who had been converted to positivism through the teachings of Alain and who had almost rejected the teaching of both Alain and Auguste Comte when the war of 1939 had broken out. False prophet, who promised peace in an age of war!

12. Auguste Comte was writing at a pivotal period in colonial

history—the moment when the empires built up between the sixteenth and eighteenth centuries had almost collapsed and the empires of the nineteenth century were about to be created. The emancipation of the American colonies of Spain had been achieved, Great Britain had lost her principal colonies in North America and France in India, Canada, and Santo Domingo. However, Great Britain had retained her empire in Asia and Canada. From 1829 to 1842, while Comte was writing the *Cours de philosophie positive,* France was beginning to build her second colonial empire by the conquest of Algeria and the acquisition of bases on the coasts of Africa and Oceania. Great Britain was doing likewise, thus taking possession of New Zealand in 1840.

Here is Comte's view of the colonial system of the seventeenth and eighteenth centuries: "Without returning, of course, to the declamatory dissertations of the last century regarding the ultimate advantage or danger of this vast operation for humanity as a whole, which is a question as idle as it is insoluble, it would be interesting to observe whether it has definitively resulted in an acceleration or a retardation of the total evolution, at once negative and positive, of modern societies. In this respect it seems at first that the new major destiny thus opened to the warlike spirit on land and on sea, and the great recrudescence similarly imparted to the religious spirit, as better adapted to the civilization of backward populations, have tended directly to prolong the general duration of the military and theological regime and, as a consequence, particularly to postpone the final reorganization. But, in the first place, the overall extension which the system of human relations had thereby tended gradually to receive has necessarily improved understanding of the true philosophical nature of such regeneration by showing it as the ultimate destiny of humanity as a whole; which should reveal the radical inadequacy of that policy, pursued on so many occasions, of systematically destroying human races when one is unable to assimilate them. In the second place, by a more direct and more immediate influence, the lively new stimulus which this great European event must everywhere have imparted to industry has certainly greatly increased its social and even its political importance, so that, all things being equal, modern evolution seems to me necessarily to have undergone a real acceleration, of which, however, most persons have a very exaggerated opinion." (*Cours de philosophie positive,* Vol. VI, p. 68.)

Comte analyzes the colonial conquests of the nineteenth century as follows: "We have, it is true, . . . noted the spontaneous appearance of a dangerous sophism which there is now a tendency to justify and which would tend to prolong military activity indefinitely by attributing the successive invasions with the specious destiny of directly establishing, in the final interest of universal civilization, the material domination of the most advanced populations over those which are less advanced. In the deplorable present state of political philosophy, which permits the short-lived ascendency of any and every aberration, such a tendency is certainly a very grave matter as a source of universal perturbation. Carried to its logical conclusion, it would no doubt, after provoking the mutual oppression of nations, culminate in the various cities attacking one another according to their unequal social progress; and without going as far as that rigorous extension which should certainly always remain ideal, it is in fact some such pretext that has served as the odious justification for colonial slavery, following the incontestable superiority of the white race. But, however grave the disorders that may temporarily be provoked by a sophism of this kind, the instinct characteristic of modern sociability must certainly dispel all irrational worry which would tend to see it, if only in the immediate future, as a new source of general wars, which would be entirely incompatible with the most persistent inclinations of all civilized populations. Before the creation and dissemination of sound political philosophy, popular rectitude will no doubt have arrived at an adequate evaluation, albeit based on a confused empiricism, of this crude retrograde imitation of the great Roman policy, which we have seen, under social conditions radically opposed to those of the modern milieu, to be sure, essentially destined to curb everywhere, except in one single people, the impending growth of military life which this empty parody would, on the contrary, simultaneously stimulate in nations long engaged in an eminently peaceful activity." (*Cours de philosophie positive,* Vol. VI, pp. 237–38.)

13. The writings of Auguste Comte are full of statements asserting the anachronism of war and emphasizing the contradiction between modern society and the military and warlike phenomenon: "All truly philosophical minds must readily acknowledge with complete intellectual and moral satisfaction that the age has finally come in which serious and lasting war must utterly disappear among the elite of humanity." (*Cours de philosophie positive,* Vol. VI, p. 239.) Or again: "All the various general means of rational exploration applicable to political study have already spontaneously concurred in observing with equal decisiveness the inevitable original tendency of humanity toward a primarily military life, and at its final destination, a no less irresistible tendency toward an essentially industrial exist-

ence. Moreover, no slightly advanced intelligence henceforth refuses to acknowledge, more or less explicitly, the continuous decline of the military spirit and the gradual rise of the industrial spirit, as a twofold necessary consequence of our progressive evolution which has been, in our own time, rather judiciously evaluated by most persons responsibly engaged in political philosophy. Indeed, in an age that sees the continual manifestation, in more and more varied forms and with ever-increasing energy, even within armies, of that repugnance for the warlike life characteristic of modern societies; when, for example, the insufficient number of men drawn to the military life has everywhere become more and more obvious due to the growing necessity for compulsory recruitment, which is rarely followed by voluntary persistence; everyday experience would doubtless dispense with all direct demonstration with regard to a notion that has filtered so gradually into the public domain. In spite of the vast exceptional development of military activity momentarily caused, at the beginning of this century, by the inevitable consequences of irresistible abnormal circumstances, our industrial and peaceful instinct soon resumed and, indeed, accelerated the regular course of its preponderant development, thus truly assuring the fundamental tranquillity of the civilized world, although the harmony of Europe must frequently seem compromised by the temporary lack of any systematic organization of international relations; which lack, without actually being capable of producing war, nevertheless often suffices to inspire dangerous anxieties. . . . Whereas industrial activity spontaneously presents that admirable ability to be stimulated simultaneously in all individuals and all peoples without giving rise to conflict, it is clear, on the contrary, that an expansion of military life in a considerable segment of humanity assumes and finally determines in the rest an inevitable compression, which constitutes the principal social function of such a regime when one considers the civilized world as a whole. Also, whereas the industrial age has no general conclusion except that still indeterminate end assigned to the progressive existence of our species by the system of natural laws, the military age must, by all necessity, have been essentially limited to the period of a sufficient gradual completion of the preliminary conditions which it was destined to fulfill." (*Cours de philosophie positive,* Vol. IV, pp. 375 and 379.)

14. "Our material riches can change hands either freely or by force. In the first case the transmission is sometimes gratuitous, sometimes interested. Similarly, involuntary transfer may be either violent or legal. These, in the last analysis, are the four general modes whereby material products are naturally

transmitted. . . . In terms of decreasing dignity and effectiveness, they must be listed in this normal order: gift, trade, inheritance, and conquest. The two middle modes are the only ones that have become very common in modern populations, as those best adapted to the industrial existence which was to prevail among them. But the two extreme modes were more effective in the initial formation of great capital. Although the last mode must ultimately fall into total disuse, this will never be the case with the first, whose importance and purity our modern industrial egotism causes us to underestimate. . . . Systematized by positivism, the impulse to give must, by making wealth both more useful and more respected, contribute to the final regime one of the best temporal auxiliaries to the continuous action of the true spiritual power. The most ancient and noble of all modes of material transmission will provide a better support for our industrial organization than can be indicated by the empty metaphysics of our crude economists." (*Système de politique positive,* Vol. II, pp. 155–56.)

This passage invites comparison with certain modern analyses, especially François Perroux's "Le don," sa signification économique dans le capitalisme contemporain," *Diogène,* April 1954, an article reprinted in *L'Économie du XXe siècle,* First edition, Paris, Presses Universitaires Françaises, 1961, pp. 322–44.

15. Michael Young, *The Rise of Meritocracy,* London, Thames and Hudson, 1958; Penguin Books, 1961.

16. Condorcet, *Esquisse d'un tableau historique des progrès de l'esprit humain.* This work, which was written in 1793, was published for the first time in 1796, the third year of the Republic. For a modern edition, see the one published by the Bibliothèque de Philosophie, Paris, Boivin, 1933. Before Condorcet, Turgot had written a *Tableau philosophique des progrès successifs de l'esprit humain.*

17. "Strictly speaking, theological philosophy, even in our early individual or social infancy, has never succeeded in being rigorously universal, which is to say that for all orders of phenomena, the simplest and commonest facts have always been regarded as essentially subject to natural laws, instead of being attributed to the arbitrary will of supernatural agents. The illustrious Adam Smith, for example, remarked very brilliantly in his philosophical essays that never in any time or place had there been found a god of gravity. This is the case, in general, even in relation to the most complicated subjects, with all phenomena sufficiently elementary and familiar that the perfect invariability of their actual relations must always have spon-

taneously impressed the least prepared observer." (*Cours de philosophie positive*, Vol. IV, p. 365.)

18. Here is another: "Despite the inevitable interdependence which, according to the principles already established, always characterizes the different elements of our social evolution, it is also necessary that, in the midst of their continuous interaction, one of these general orders of progress be spontaneously dominant so that it habitually imparts to all the others an indispensable primitive impulse, although it in turn must subsequently receive a new impulse from their own evolution. It suffices here to discern immediately this dominant element, whose consideration must guide the whole of our dynamic exposition, without concerning ourselves explicitly with the particular subordination of the other elements to it or among themselves, which will be adequately revealed by the spontaneous execution of such a project. Now, reduced to these terms, the determination of this element can offer no serious difficulty, since one need only distinguish that social element whose development could best be conceived apart from the development of all the others, in spite of their universal and necessary relatedness; whereas this notion would, on the contrary, inevitably multiply in the direct consideration of the development of these others. By virtue of this doubly decisive quality, no one could hesitate to award first place to intellectual evolution as the necessarily dominant principle of the whole evolution of humanity. If, as I have explained in the preceding chapter, the intellectual point of view must predominate in the simple static study of the social organism properly speaking, there is all the more reason that this should be so in the direct study of the general movement of human societies. Although our feeble intelligence undoubtedly has an indispensable need for the original awakening and continuous stimulation imparted by the appetites, the passions, and the sentiments, it is nevertheless under its necessary direction that the whole of human progress has always been accomplished. . . . And in all times, from the first flight of philosophical genius, the history of society has always been acknowledged more or less distinctly but always unimpeachably to be primarily dominated by the history of the human mind." (*Cours de philosophie positive*, Vol. IV, pp. 340–42.)

19. Thus, in the *Discours sur l'esprit positif*, Comte writes: "Polytheism was adapted primarily to the system of conquest of antiquity, and monotheism to the defensive organization of the Middle Ages. In giving greater and greater predominance to industrial life, modern sociability must therefore lend powerful support to the great mental revolution which today is definitively elevating our intelligence from the theological regime to

the positive regime. Not only is this active daily tendency toward the practical improvement of the human condition necessarily incompatible with religious preoccupations, which are always concerned, especially under monotheism, with a completely different destination; but also such an activity must by its very nature ultimately give rise to a universal opposition, as radical as it is spontaneous, to all theological philosophy." (Ed. 10/18, Paris, Union Générale d'Editions, 1963, pp. 62–63.)

20. Oswald Spengler, *Der Untergang des Abendlandes—Umrisse einer Morphologie des Weltgeschichte,* Munich, 1918–1922. This book, which was conceived at the time of the Agadir incident, was first published in 1916. But its success, which was spectacular in Germany, did not come until after the defeat of 1918.

21. "The three general sources of social variation seem to me to be: 1) race; 2) climate; and 3) political action properly speaking, considered in its full scientific extension. This is not the place to take up the question of whether this order truly corresponds to their relative importance. Although this consideration obviously would not be inappropriate in the nascent state of the science, the laws of method oblige us to defer its direct exposition at least until after examination of the principal subject, in order to avoid an irrational confusion between the fundamental phenomena and their various modifications." (*Cours de philosophie positive,* Vol. IV, p. 210.)

22. After raising the question, at the beginning of the fifty-second lecture of the *Cours de philosophie positive,* "why the white race possesses to such a marked degree the actual privilege of principal social development and why Europe has been the essential site of this dominant civilization," Comte, after stating that "this great discussion of concrete sociology" must "be reserved until after the first abstract formulation of the fundamental laws of social development," nevertheless give several reasons, "necessarily inadequate partial and isolated glimpses:" "As to the first question, no doubt one already perceives in the organization characteristic of the white race, especially in terms of the cerebral apparatus, a few positive germs of its real superiority; however, naturalists are still very far from a satisfactory agreement on this matter. Similarly, as for the second observation, one can perceive in a somewhat more satisfactory manner various physical, chemical, and even biological conditions which must certainly have influenced to some degree the eminent right of the European countries to serve hitherto as the main theater of this dominant evolution of humanity." And Comte particularizes:

"Such are, for example, from the physical point of view, besides the thermologically advantageous situation of the temperate zone, the existence of the admirable Mediterranean basin, around which in the beginning the most rapid social development must have occurred once the art of navigation had become sufficiently advanced to permit the use of this precious intermediary, which offered all the coastal nations both the contiguity calculated to facilitate sustained relations, and the diversity which made these relations important to mutual social stimulation. Similarly, from the chemical point of view, the greater abundance of iron and coal in these privileged countries must certainly have contributed a great deal to the acceleration of the inhuman evolution. Finally, from the biological point of view, whether phytological or zoological, it is clear that since this same milieu was more favorable not only to the principal food crops, but also to the development of the most valuable domestic animals, civilization must also, if for this reason alone, have been especially encouraged. But whatever real importance one can now attach to these various suggestions, they are obviously still far from adequate to provide a truly positive explanation for the proposed phenomenon. When the appropriate formation of the social dynamic will subsequently have made it possible to go ahead with such an explanation, it is still evident that each of the preceding indications will first need to be subjected to a scrupulous scientific revision based on the whole of natural philosophy." (*Cours de philosophie positive,* Vol. V, pp. 12–13.)

23. Auguste Comte is extremely severe with Napoleon: "By a fatality that will ever be deplorable, the inevitable military supremacy to which the great Hoche seemed at first so happily destined fell to a man who was almost a stranger to France, sprung from a backward civilization, and specially motivated, under the secret impulsion of a superstitious nature, by an involuntary admiration for the ancient social hierarchy; whereas the boundless ambition with which he was consumed did not, in spite of his vast characteristic charlatanism, truly correspond to any eminent mental superiority save that associated with an undeniable talent for war which is much more closely related, especially in our time, to moral energy than to intellectual force.

"One cannot recall his name today without remembering that base flatterers and ignorant enthusiasts dared for a long time to compare with Charlemagne a sovereign who in all respects was just as much behind his age as that admirable model of the Middle Ages had been ahead of his. Although all personal evaluation must remain essentially alien to the nature and purpose of our historical analysis, every true philosopher must, in my opinion, now regard it as his inescapable social duty properly to acquaint the public reason with that dangerous aberration which, under the mendacious leadership of a press as guilty as it is misguided, is today driving the whole revolutionary school to try, through a fatal blindness, to rehabilitate the memory, at first so justly abhorred, of him who organized in the most disastrous manner the gravest political retrogression that humanity has ever suffered." (*Cours de philosophie positive*, Vol. VI, p. 210.)

24. The influence of positivism was very profound in Brazil, where it even became the quasi-official doctrine of the state. Benjamin Constant, President of the Republic, gave the *Encyclopédie des sciences positives* founded by Comte as a course of study in the public schools. An Institute of the Apostolate was founded in 1880 and a positivist temple established in Rio in 1891 to observe the Cult of Humanity there. The motto "Order and progress" (*Ordem e Progresso*) appears on the green flag of Brazil. Green was also the color of the positivist flags.

25. The distinction between the reason and the heart is found in Plato, in *The Republic* and in *Phaedrus*. It reappears in a physiological description of mortal living creatures in *Timaeus* (#69 ff.) in which Plato presents a table of the corporal localizations, situating the immortal soul in the head and the mortal soul in the chest. There are other similarities between the thought of Plato and Comte. For instance, Plato's myth of the charioteer (cf. *Phaedrus*) is not unlike the dialectic which Comte discovers in man between affection, action, and intelligence.

26. "From this social point of view, the institution of language must ultimately be compared with that of property. . . . For the former performs for the spiritual life of humanity a basic function which corresponds to the one the latter fulfills with respect to its material life. After essentially facilitating the acquisition of all of human knowledge, theoretical or practical, and guiding our aesthetic impulse, language consolidates this twofold wealth and transmits it to new partners. But the diversity of the deposits establishes a fundamental difference between the two conservative institutions. For productions intended to satisfy personal needs which necessarily destroy them, property must establish individual guardians whose social effectiveness is even increased by a prudent concentration. On the contrary, in the case of wealth that allows for simultaneous possession without suffering any alteration, language naturally establishes a full community where all, by drawing freely upon

the universal treasure, spontaneously cooperate in its preservation. In spite of this fundamental difference, the two systems of accumulation give rise to comparable abuses, resulting from the desire to enjoy without producing. The guardians of material goods may degenerate into exclusive arbiters of their use, which is too often directed toward selfish satisfaction. Similarly, those who have really contributed nothing to the spiritual treasury deck themselves out in it and thus usurp a brilliance which excuses them from any real service." (*Système de politique positive*, Vol. II, p. 254.)

27. For Auguste Comte there is only *one* history of man and it is his ambition to integrate all moments of the past into his synthesis. Even in this sense of tradition he finds one of the principal superiorities of positivism: "The Western anarchy consists primarily in the alteration of the continuity of mankind, which has been successively violated by Catholicism, which condemned antiquity, Protestantism, which disapproved of the Middle Ages, and deism, which denied all filiation. Nothing could better express the need for positivism to provide at last the only possible solution to the revolutionary situation by transcending all these more or less subversive doctrines which have gradually driven the living to rebel against the company of the dead. After a service of this kind history will soon become the sacred science, according to its normal function, the direct study of the destinies of that Great Being whose notion includes all our sound theories. Systematized politics will henceforth make this notion the ground of all its enterprises, which will be naturally subordinate to the corresponding stage of the great evolution. Even poetry, regenerated, will draw from this source the images destined to prepare the future by idealizing the past." (*Système de politique positive*, Vol. III, p. 2.)

28. "So it is that the only principle of cooperation on which political society properly speaking is based naturally gives rise to the government that must maintain and develop this society. Such a power appears, in truth, as essentially material, since it always results from number or wealth. But it is important to recognize that the social order can never have any other immediate foundation. Hobbes's celebrated principle of the natural domination of force really constitutes the only major advance in the positive theory of government between Aristotle and myself. For the admirable anticipation, in the Middle Ages, of the separation of the two powers was, in a favorable situation, due more to sentiment than to reason: the idea was later unable to withstand debate until I revived it early in my career. All the odious reproaches incurred by Hobbes's conception resulted solely from its metaphysical source and from the resulting radi-

cal confusion between static and dynamic valuation, which people were unable to distinguish at the time. But given less spiteful and more enlightened judges, this double imperfection would have served only to emphasize the difficulty as well as the importance of this luminous insight, of which only positive philosophy could make sufficient utilization." (*Système de politique positive*, Vol. II, p. 299.)

29. "But the habitual harmony between functions and functionaries will always show great imperfections. Although one would like to see everyone in his place, the short duration of our objective life necessarily prevents us from reaching this state, because we are unable to examine the titles closely enough to bring about the changes in time. Moreover, it must be recognized that most social offices require no truly natural aptitude which cannot be fully compensated by an appropriate training, from which nothing can completely excuse us. Since the best agent always has need of a special apprenticeship, one must always have considerable respect for every actual possession of functions as well as of capital, recognizing how important this personal security is to social effectiveness. Besides, we should pride ourselves less on natural qualities than on acquired advantages, since we have had less to do with the former. Therefore, our true merit, like our happiness, depends above all upon the worthy voluntary use of whatever forces the existing order, artificial as well as natural, makes available to us. Such is the wholesome view according to which the spiritual power must continually inspire individuals and classes with a prudent resignation toward the necessary imperfections in the social harmony, whose greater complexity exposes it to more abuses.

"This habitual conviction would, however, be inadequate to restrain anarchical demands if the underlying sentiment did not at the same time receive a certain normal satisfaction, suitably regulated by the priesthood. This satisfaction results from the evaluative aptitude which constitutes the principal characteristic of spiritual power, from which all its social functions of counsel, consecration, and discipline obviously derive. Now, this evaluation, which necessarily begins with offices, must ultimately extend as far as individual agents. The priesthood must, no doubt, always attempt to curb personal alterations, whose free play would soon become more baneful than the abuses which have inspired them. But it must also create and develop, in contrast to this objective order resulting from actual power, a subjective order based on personal esteem, according to an adequate evaluation of all individual rights. Although this second system of classification cannot and must not ever prevail, except in sacred worship, its just opposition to the first gives rise to the really practicable improvements even as it consoles for the ineradicable imperfections." (*Système de politique positive,* Vol. II, pp. 329–30.)

30. "True philosophy attempts, insofar as possible, to systematize all of human existence, individual and above all collective, considered simultaneously in terms of the three order of phenomena which characterize it: thoughts, sentiments, and acts. From all these aspects, the fundamental evolution of humanity is necessarily spontaneous, and only a correct understanding of its natural course can provide us with a general basis for a prudent intervention. But the systematic modifications that we can introduce are nevertheless extremely important in that they greatly diminish the partial deviations, harmful retardations, and grave inconsistencies that would naturally attend such a complex movement if it were left entirely to itself. The continuous realization of this indispensable intervention constitutes the essential domain of politics. Nevertheless, its true conception can only emanate from philosophy, which incessantly perfects its general determination. With respect to this common fundamental destination, the proper function of philosophy consists in coordinating all parts of human existence in order to reduce its theoretical notion to a complete unity. A synthesis of this kind can be true only insofar as it accurately represents the totality of natural relations, the judicious study of which thus becomes the preliminary condition for its creation. If philosophy tried directly to influence active existence save by this systematization, it would be falsely usurping the necessary mission of politics, which is the sole legitimate arbiter of all practical evolution. Between these two principal functions of the great organism, the enduring bond and the normal separation both reside in systematic ethics, which naturally constitutes the proper application of philosophy and the general guide of politics." (*Système de politique positive,* Vol. I, *Discours préliminaire,* p. 8.)

31. Comte expounds his philosophy of knowledge in the chapter on the religion of social statics in the *Système de politique positive:*

"Sound philosophy . . . regards all real laws as constructed by us from external materials. Evaluated objectively, their accuracy can never be anything but approximate. But since they are created only for our needs, especially our active needs, these approximations become quite sufficient when they are well established according to the practical requirements which habitually determine appropriate precision. Beyond this principal standard there often remains a normal degree of theoretical freedom. . . .

"Our fundamental construction of the universal order results, therefore, from a necessary union of the outside and the inside. The real laws, that is, the general facts, are never anything but hypotheses sufficiently confirmed by observation. If no harmony existed outside of us, our minds would be entirely incapable of conceiving it; but in no case is it verified to the extent that we suppose. In this continuous cooperation, the world provides the substance and man the form of each positive notion. But the fusion of these two elements becomes possible only by mutual sacrifice. An excess of objectivity would prevent any general view, which is always based on abstraction. But the decomposition which permits us to abstract would be impossible if we did not overcome a natural excess of subjectivity. Every man, in comparing himself to others, spontaneously purges his own observations of what is originally too personal about them in the interests of that social harmony which constitutes the principal end of contemplative life. But the degree of subjectivity which is common to our species ordinarily persists, without any grave disadvantage, for that matter. . . .

"If [the universal order] were completely objective or purely subjective it would long since have been grasped by our observations or have emanated from our conceptions. But the notion of this universal order requires the union of two heterogeneous, albeit inseparable, influences whose combination could only develop very slowly. The various irreducible laws which make up this order form a natural hierarchy in which each category is based on the preceding category according to their decreasing generality and increasing complexity. Thus their sound evaluation has necessarily been gradual." (*Système de politique positive*, Vol. II, p. 32–34.)

32. The fifteen laws of the primary philosophy are set forth in Volume IV of the *Système de politique positive* (Chapter 3, pp. 173–81).

33. "The Great Being is the totality of beings past, future, and present, which freely unite to perfect the universal order." (*Système de politique positive*, Vol. IV, p. 30.)

"The worship of truly superior men forms an essential part of the worship of Humanity. Even during his objective lifetime, each constitutes a certain personification of the Great Being. However, this representation requires that one set aside ideally the grave imperfections which often alter the best natural men." (*Ibid.*, Vol. II, p. 63.)

"Not only is Humanity composed only of existences susceptible to assimilation, but of each of these existences it admits only the incorporable part, overlooking every individual deviation." (*Ibid.*, Vol. II, p. 62.)

Karl Marx

I

IT IS REALLY no more difficult to present Marx's leading ideas than those of Montesquieu or Comte; if only there were not so many millions of Marxists, there would be no question at all about what Marx's leading ideas are or what is central to his thought.

Marx was, first and foremost, the sociologist and economist of the capitalist regime. He had a certain conception of that regime, of the destiny it imposed upon men, and of the evolution it would undergo. As sociologist-economist of the regime he called capitalist, he had no precise image of what the socialist regime would be, and he repeatedly said that man cannot know the future in advance. It is of no great interest, therefore, to wonder whether, if Marx were alive, he would be a Stalinist, a Trotskyite, a Khrushchevian, a Maoist, or something else. Marx had the fortune, or misfortune, to live a century ago. He gave no answers to the questions we are asking today. We can give those answers for him, but they are our answers and not his. To wonder what Marx would have thought, had he lived in another century, is like wondering what another Marx would have thought, instead of the real Marx. An answer is possible, but it is problematical and of questionable interest.

Nevertheless, it is certainly a complicating fact that today almost a billion human beings are taught a doctrine which, rightly or wrongly, is labeled Marxist. A certain interpretation of Marx's doctrine has become the official ideology first of the Russian state, next of the states of Eastern Europe, and finally of the Chinese state. This official doctrine claims to give the true interpretation of Marx's thought. It is enough for a sociologist or, more modestly, a teacher to present a certain interpretation of Marx's thought for him to become, in the eyes of supporters of the official doctrine, a mouthpiece of the bourgeoisie, of capitalism, and of imperialism.

The official doctrine presents qualities of oversimplification and exaggeration inseparable from the fact of its being taught as a catechism to mentalities of various sorts. However, there are also—outside the dominion of official Marxism—self-proclaimed Marxist thinkers who provide us with a series of interpretations, each more intelligent and ingenious than the last, of Marx's innermost, ultimate thought. These interpretations have given rise to passionate debates, interesting publications, and learned controversies —all of which, however, belong more to café philosophy than to world history.

I shall not strive for a supremely ingenious interpretation of Marx. I believe that Marx's central ideas are simpler than many Marxists would have us think and that they are to be found, not in his youthful or marginal writings, but in those which he published and which he himself always regarded as the chief expression of his thought.

Not that there are no intrinsic difficulties. These difficulties have to do, first, with the fact that Marx was a prolific writer and that, as is sometimes the case with sociologists, he wrote both journalism and immense books of several hundred pages. Since he wrote a great deal, he did not always say the same thing on the same subject. With a little ingenuity and erudition, one can find Marxist formulas on most subjects which do not seem to agree, or which at least lend themselves to various interpretations.

Moreover, Marx's canon includes works of sociological theory, economic theory, and history, and sometimes the explicit theory to be found in the scientific writings seems to be contradicted by the implicit theory employed in the books of history. For example, Marx offered a certain conception of class; but when he analyzed historically the class struggle in France between 1848 and 1850 or the *coup d'état* of Napoleon III or the history of the Commune, the classes which he recognized and to which he assigned roles in the drama are not necessarily those implied by his theory.

Besides the diversity of works, we must take into consideration the diversity of periods. There has been a general agreement to distinguish two main divisions. The first, the so-called youthful period, comprises the writings between 1841 and 1847 or 1848. Among the writings of this stage, some were published during Marx's lifetime: short articles or essays like the *Introduction to the Critique of Hegel's Philosophy of Law* or the essay on *The Jewish Question*.

Among the more important works of this period which were known for some time are *The Holy Family* and a polemic against Proudhon entitled *The Poverty of Philosophy*, a reply to Proudhon's book *The Philosophy of Poverty*. The other writings of this period were not published until long after Marx's death, because publication of the whole body of his work dates from 1931. It is only after this date that there arose a whole literature reinterpreting Marx's thought in the light of his youthful writings. Among these writings that have received considerable attention, there are a fragment of a critique of Hegel's philosophy of law, a text entitled *Economic and Philosophic Manuscripts of 1844*, and finally a very important work entitled *The German Ideology*.

This youthful period is thought to end with *The German Ideology*, *The Poverty of Philosophy*, and, above all, the famous little classic called *The Communist Manifesto*, a masterpiece of the sociological literature of propaganda in which Marx's leading ideas are set forth for the first time, in a manner both lucid and impressive.

It might be said that, from 1848 until the end of his life, Marx apparently ceased to be a philosopher and became a sociologist and, above all, an economist. The majority of those who now declare themselves to be more or less Marxists have the annoying peculiarity of being ignorant of the political economy of our age; Marx did not share this weakness. He had received an admirable economic education and knew the economic thinking of his time as few men did. He was, and wanted to be, an economist in the strict and scientific sense of the word.

In this second period of his life, his two most important works are an 1859 text entitled *A Contribution to the Critique of Political Economy* and, of course, Marx's masterpiece, the heart of his thought, *Capital.*

Marx had a certain philosophical vision of the historical process. That he gave the contradictions of capitalism a philosophical significance is possible and even probable. But the essence of Marx's scientific effort has been to demonstrate scientifically what was for him the inevitable evolution of the capitalist system. Any interpretation of Marx which finds no place for *Capital,* or is able to summarize it in a few pages, is a deviation from what Marx himself thought and desired.

It is always possible to say that the great thinker misunderstood himself and that the essential texts are those he scorned to publish. But one must be very sure of one's genius to be convinced of understanding a great writer so much better than he understood himself. When one is less sure of one's genius, it is better to begin by understanding the writer as he understood himself and thus to assign the central place in Marxism to *Capital,* and not to the *Economic and Philosophic Manuscripts* or *The German Ideology,* the incomplete (though perhaps highly original) rough drafts of a young man speculating on Hegel and capitalism at a time when he certainly knew Hegel better than he knew capitalism.

I shall therefore begin my own analysis of Marx with Marx's mature thought, i.e., with 1848, and it is in *The Communist Manifesto,* the *Contribution to the Critique of Political Economy,* and *Capital* that I shall seek Marx's own thought; I shall reserve for later an inquiry into the philosophical background of his historico-sociological concepts.

One last remark, before beginning my exposition. For over a century, many schools shared a tendency to call themselves Marxist, although they offered different versions of his thought. In my exposition, I shall not attempt to judge their merit. Instead, I shall try to show why the themes of Marx's thought are simple and deceptively clear but lend themselves to interpretations among which it is impossible to choose with certainty. Any theory that becomes the ideology of a political movement or the official doctrine of a state must lend itself to simplification for the simple and to subtlety for the subtle. There is no question that Marx's thought presents these virtues in the highest degree.

As I have said, Marx's thought is an analysis and an interpretation of capitalistic society in terms of its current functioning, its present structure, and its necessary evolution. Auguste Comte had developed a theory of what he called industrial society, that is, of the major characteristics of all modern societies. The essential antithesis in Comte's thinking is between the feudal, military, theological societies of the past and the industrial and scientific societies of the present. Unquestionably Marx, too, believed that modern societies are industrial and scientific in comparison with past military and theological societies. But instead of assigning the central position in his interpretation to the antimony between the societies of the past and the societies of the present, Marx assigned it to the contradiction, in his eyes inherent in modern society, which he called capitalism.

While in positivism the conflicts between labor and management are marginal phenomena, imperfections of industrial society which are relatively easy to correct, in Marx's thought conflicts between labor and management—or, to use the Marxist vocabulary, between the proletariat and the capitalists—are the major fact of modern societies, the one that reveals the essential nature of these societies and thereby enables us to anticipate their historical development. Marx's thought is an interpretation of the contradictory or antagonistic character of capitalist society. In a certain sense, Marx's whole canon is an attempt to show that this antagonistic character is inseparable from the fundamental structure of the capitalist system and is, at the same time, the mechanism of the historic movement.

I shall present rapidly the major ideas first of *The Communist Manifesto,* next of the Preface to *A Contribution to the Critique of Political Economy,* and finally of *Capital.* These three famous texts are three ways of explaining, establishing, and describing the antagonistic character of the capitalist system.

If it is clearly understood that the center of Marx's thought is his assertion of the antagonistic character of the capitalist system, then it is immediately apparent why it is impossible to separate the analyst of capitalism from the prophet of socialism or, again, the sociologist from the man of action; for to show the antagonistic character of the capitalist system irresistibly leads to predicting the self-destruction of capitalism and thence to urging men to contribute something to the fulfillment of this prearranged destiny.

The Communist Manifesto is a propaganda pamphlet in which Marx and Engels presented some of their scientific ideas in collective form. Its central theme is the class struggle. All history is the history of the class struggle: free men and slaves, patricians and plebeians, barons and serfs, master artisans and journeymen—in short, oppressors and oppressed—have been in constant opposition to one another and have carried on an unceasing struggle, at times secret, at times open, which has always ended with a revolutionary transformation of the whole society or with the mutual destruction of the warring classes.

Here, then, is the first decisive idea: human history is characterized by the struggle of human groups which will be called social classes, whose definition remains for a moment ambiguous, but which are characterized in the first place by an antagonism between oppressors and oppressed and in the second place by a tendency toward a polarization into two blocs, and only two.

All societies having been divided into warring classes, modern capitalist society does not differ from those that preceded it. But the ruling and exploiting class of modern society, namely the bourgeoisie, presents certain characteristics which are without precedent.

The bourgeoisie is incapable of maintaining its ascendancy without permanently revolutionizing the instruments of production. The bourgeoisie, said Marx, has developed the forces of production further in a few decades than previous societies have done in many centuries.[1] Engaged in merciless competition, the capitalists are incapable of *not* revolutionizing the means of production. The bourgeoisie is creating a world market; it is destroying the remnants of the feudal system and the traditional communities. But just as the forces of production which gave rise to the capitalist regime had developed in the heart of feudal society, so the forces of production which will give rise to the socialist regime are ripening in the heart of modern society.

What is the basis of this antagonism characteristic of capitalist society? It is the contradiction between the forces and the relations of production. The bourgeoisie is constantly creating more powerful means of production. But the relations of production—that is, apparently, both the relations of ownership and the distribution of income—are not transformed at the same rate. The capitalist system is able to produce more and more, but in spite of this increase in wealth, poverty remains the lot of the majority. This contradiction will eventually produce a revolutionary crisis. The proletariat, which constitutes and will increasingly constitute the vast majority of the population, will become a class, that is, a social entity aspiring to the seizure of power and the transformation of social relations. But the revolution of the proletariat will differ in kind from all past revolutions. All the revolutions of the past were accomplished by minorities for the benefit of minorities. The revolution of the proletariat will be accomplished by the vast majority for the benefit of all. The proletarian revolution will therefore mark the end of classes and of the antagonistic character of capitalist society.

Marx did not deny that between capitalists and proletarians there are today a number of intermediate groups—artisans, *petite bourgeoisie,* merchants, peasant landowners. But he made two statements. First, along with the evolution of the capitalist regime there will be a tendency toward crystallization of social relations into two groups, and only two: the capitalists on the one hand, and the proletarians on the other. Two classes, and only two, represent a possibility for a political regime and an idea of a social regime. On the day of the decisive conflict, every man will be obliged to join either the capitalists or the proletarians. On the day when the proletarian class seizes power, there will be a final break with the course of all previous history. In fact, the antagonistic character of all known societies will disappear.

We read at the end of the second chapter of *The Communist Manifesto:*

> When, in the course of development, class distinctions have disappeared and all production has been concentrated in the hands of a vast association of the whole nation, the public power will lose its political character. Political power, properly so called, is merely the organized power of one class for oppressing another. If the proletariat during its contact with the bourgeoisie is compelled by the force of circumstances to organize itself into a class; if by means of a revolution it makes itself into the ruling class and, as such, sweeps away by force the old conditions of production, then it will, along with these conditions, have swept away the conditions for the existence of class antagonisms and of classes generally, and will thereby have abolished its own supremacy as a class.
>
> In place of the old bourgeois society, with its classes and class antagonisms, we shall have an association in

which the free development of each is the condition
for the free development of all.

This passage is altogether characteristic of one of the
essential themes of Marx's theory. We have seen from
Comte's positivism that the writers of the early nineteenth
century had a tendency to regard politics or the state as a
phenomenon secondary to the essential, economic or social,
phenomena. Marx belonged to this general trend, and he,
too, regarded politics and the state as phenomena secondary
to what is happening within the society itself.

As a consequence, he presented political power as the
expression of social conflicts. Political power is the means
by which the ruling class, the exploiting class, maintains its
domination and its exploitation. According to this line of
thought, the abolition of class contradictions must logically
entail the disappearance of politics and of the state, because
politics and the state are seemingly the by-products or the
expressions of social conflicts.

Such are the themes of Marx's historical vision and of
his political propaganda. The aim of his science is to provide a strict demonstration of the antagonistic character of
capitalist society, the inevitable self-destruction of an antagonistic society, and the revolutionary explosion that will
put an end to the antagonistic character of modern society.

II

I HAVE SUGGESTED that the center of Marx's thought was his
interpretation of the capitalist system as contradictory, i.e.,
as dominated by the class struggle. Auguste Comte thought
that the society of his age lacked consensus because of the
juxtaposition of institutions dating from theological and
feudal societies and institutions belonging to industrial society. Observing the lack of consensus about him, he turned
to the past for a demonstration of the principles of consensus in historical societies. Marx observed—or thought he

observed—a class struggle in capitalist society, and he re-discovered in the different societies of history the equivalent of the class struggle.

According to Marx, the class struggle tends toward a simplification. The different social groups will be polarized, some around the bourgeoisie, others around the proletariat, and it is the development of the forces of production that will be the mechanism of the historical movement. This movement will result, through proletarization and pauperization, in a revolutionary explosion and the advent, for the first time in history, of a society without conflict.

Starting with these general themes of Marx's historical interpretation, we have two questions to answer. First, what is Marx's general theory of society which accounts both for the contradictions of present society and for the antagonistic character of all known societies? Second, what are the structure, the mode of operation, and the specific evolution of capitalist society which explain the modern class struggle and the revolutionary outcome of the capitalist system?

In other words, beginning with the Marxist themes we find in *The Communist Manifesto,* we must explain: (1) the general theory of society commonly called *historical materialism* or, more recently, *dialectical materialism* and (2) Marx's essential economic ideas as they are found in *Capital.*

To simplify the exposition, I shall quote a famous passage —probably the most famous in Marx—which occurs in the Preface to *A Contribution to the Critique of Political Economy,* in which Marx summarized his sociological conception as a whole:

> The general conclusion at which I arrived and which, once obtained, served to guide me in my studies, may be summarized as follows. In the social production which men carry on they enter into definite relations that are indispensable and independent of their will; these relations of production correspond to a definite stage of development of their material pow-

ers of production. The sum total of these relations of production constitutes the economic structure of society—the real foundation on which rise legal and political superstructures and to which correspond definite forms of social consciousness. The mode of production in material life determines the general character of the social, political and spiritual processes of life. It is not the consciousness of men that determines their existence, but, on the contrary, their social existence determines their consciousness. At a certain stage of their development, the material forces of production in society come into conflict with the existing relations of production, or—what is but a legal expression for the same thing—with the property relations within which they had been at work. From forms of development of the forces of production, these relations turn into their fetters. Then comes the period of social revolution. With the change of the economic foundation the entire immense superstructure is more or less rapidly transformed. In considering such transformation the distinction should always be made between the material transformation of the economic conditions of production which can be determined with the precision of natural science, and the legal, political, religious, aesthetic or philosophic—in short ideological—forms in which men become conscious of this conflict and fight it out. Just as our opinion of an individual is not based on what he thinks of himself, so we cannot judge of such a period of transformation by its own consciousness; on the contrary, this consciousness must rather be explained from the contradictions of material life, from the existing conflict between the social forces of production and the relations of production. No social order ever disappears before all the productive forces, for which there is room in it, have been developed; and the new higher relations of production never appear before the material conditions of their existence have matured in the womb of the old society. Therefore, mankind always takes up only such problems as it can solve; since, looking at the matter more closely, we will always find that the problem itself arises only when the material conditions necessary for its solution already exist or are at least in the process of formation. In broad outlines we can designate the Asiatic, the ancient, the feudal, and the modern bourgeois methods of production as so many epochs in the progress of the economic formation of society. The bourgeois relations of production are the last antagonistic form of the social process of production—antagonistic not in the sense of individual antagonism, but of one arising from conditions surrounding the life of individuals in society; at the same time the productive forces developing in the womb of bourgeois society create the material conditions for the solution of that antagonism. This social formation constitutes, therefore, the closing chapter of the prehistoric stage of human society.

This passage contains all the essential ideas of Marx's economic interpretation of history—with the sole reservation, to which I call attention, that neither the concept of class nor the concept of class struggle figures in it explicitly.

The first and essential idea is that men enter into definite relations that are independent of their will. In other words, we can follow the movement of history by analyzing the structure of societies, the forces of production, and the relations of production, and not by basing our interpretation on men's ways of thinking about themselves. There are social relations which impose themselves on individuals exclusive of their preferences, and an understanding of the historical process depends on our awareness of these supraindividual social relations.

Second, in every society there can be distinguished the economic base, or infrastructure, as it has come to be called, and the superstructure. The infrastructure consists essentially of the forces and relations of production, while within the superstructure figure the legal and political institutions as well as ways of thinking, ideologies, and philosophies.

Third, the mechanism of the historical movement is the contradiction, at certain moments in evolution, between the forces and the relations of production. The forces of production seem to be essentially a given society's capacity to produce, a capacity which is a function of scientific knowledge, technological equipment, and the organization of collective labor. The relations of production, which are not too precisely defined in the passage I have quoted, seem to be essentially distinguished by relations of property. Indeed, we have the formula, "existing relations of production, or —what is but a legal expression for the same thing— . . . the property relations within which they had been at work." However, relations of production need not be identified with relations of property; or at any rate relations of production may include, in addition to property relations, distribution of national income (which is itself more or less strictly determined by property relations).

Fourth, into this contradiction between forces and relations of production it is easy to introduce the class struggle, although, again, the passage I have quoted does not refer to it. Indeed, we need merely suppose that in revolutionary periods—that is, periods of contradiction between the forces and relations of production—one class is attached to the old relations of production which are becoming an obstacle to the development of the forces of production, and another class, on the contrary, is progressive and represents new relations of production which, instead of being an obstacle in the way of the development of the forces of production, will favor the maximum growth of those forces.

If we turn from these abstract formulas to the interpretation of capitalism, the result is as follows: In capitalist society, the bourgeoisie is attached to private ownership of the means of production and therefore to a certain distribution of national income. On the other hand, the proletariat, which constitutes the opposite pole of society and represents another organization of the collectivity, becomes, at a certain moment in history, the representative of a new social organization which will be more progressive than the capitalist organization. This new organization will mark a later phase of the historical process, a further development of the forces of production, a stage in the course of a progressive history.

Fifth, this dialectic of the forces and relations of production also implies a theory of revolution. For in this vision of history, revolutions are not political accidents, but the expression of a historical necessity. Revolutions perform necessary functions. Revolutions occur when the conditions for them are given. Remember the key sentence: "No social order ever disappears before all the productive forces for which there is room in it have been developed; and the new higher relations of production never appear before the material conditions of their existence have matured in the womb of the old society."

Capitalist relations of production were first developed in the womb of feudal society. The French Revolution occurred when the new capitalist relations of production had attained a certain degree of maturity. And, at least in this passage, Marx foresaw an analogous process for the transition from capitalism to socialism. The forces of production must be developed in the womb of capitalist society; socialist relations of production must mature in the womb of the present society before the revolution which will mark the end of "prehistory" is to occur. It is because of this theory of revolution that the Second International, social democracy, tended toward a relatively passive attitude; the forces and relations of production of the future had to be allowed to mature before a revolution could be accomplished. Mankind, said Marx, always takes up only such problems as it can solve: social democracy was afraid of bringing about the revolution too soon, which explains why it never did; but that is another matter.

Sixth, in this historical interpretation, Marx not only distinguished infrastructure and superstructure; he also opposed social reality to consciousness. It is not men's consciousness that determines reality; on the contrary, it is the social reality that determines their consciousness. This results in an over-all conception in which men's ways of thinking must be explained in terms of the social relations of which they are a part. Statements of this kind may provide a basis for what is referred to today as the sociology of knowledge.

Finally, a last theme contained in this passage: Marx sketched the stages of human history. Just as Auguste Comte differentiated moments of human evolution on the basis of ways of thinking, so Marx differentiated stages of human history on the basis of their economic regimes; and he distinguished four of these or, in his terminology, four modes of production which he called the Asiatic, the ancient, the feudal, and the bourgeois.

These four modes may be divided into two groups. The ancient, feudal, and bourgeois modes of production have been realized in the history of the West. They are the three stages of Western history, and each is characterized by the type of relationship among the men who work. The ancient mode of production is characterized by slavery, the feudal mode of production by serfdom, and the bourgeois mode of production by wage earning. They constitute three distinct modes of man's exploitation by man. The bourgeois mode of production constitutes the last antagonistic social formation because, or rather to the extent that, the socialist mode of production, i.e., the association of producers, no longer involves man's exploitation by man or the subordination of manual laborers to a class wielding both ownership of the means of production and political power.

On the other hand, the Asiatic mode of production does not seem to constitute a stage in Western history. Hence Marx's interpreters have endlessly debated the unity or non-unity of the historical process. For if the Asiatic mode of production characterizes a civilization distinct from the West, it is probable that several lines of historical evolution are possible, depending on the human group in question. Moreover, the Asiatic mode of production does not seem to be distinguished by the subordination of slaves, serfs, or wage earners to a class possessing the instruments of production, but by the subordination of all the workers to the state. If this interpretation of the Asiatic mode of production is correct, the social structure would be characterized not by class struggle in the Western sense of the term, but by the exploitation of the whole society by the state or the bureaucratic class.

You see what use can be made of the notion of the Asiatic mode of production. Indeed, it is conceivable that in the event of the socialization of the means of production, capitalism might result, not in the end of all exploitation, but in the spread of the Asiatic mode of production to all mankind. Those sociologists who dislike Soviet society have commented at length on these passing remarks on the Asiatic mode of production. They have even found in Lenin certain passages expressing the fear that a socialist revolution might result, not in the end of man's exploitation by man, but in the Asiatic mode of production, and from the passages they have drawn conclusions whose political nature may be readily divined.[2]

But for the moment we are not concerned with these political conclusions. Rather, we must recognize the fact that Marx, considering that each society is characterized by its infrastructure or mode of production, distinguished four modes of production, or four stages in the history of mankind, prior to the socialist mode of production, which is situated beyond "prehistory."

Such, in my opinion, are the leading ideas of Marx's economic interpretation of history. We have not been concerned with such a complex philosophical problem as: To what extent is this economic interpretation separable or inseparable from a materialist metaphysic? For the moment, let us confine ourselves to the leading ideas which are obviously those Marx expounded and which, moreover, admit of a certain degree of obscurity or ambiguity, inasmuch as the exact definitions of infrastructure and superstructure may constitute—and have constituted—the subject of endless debate.

Let us now turn to the second task on our agenda, namely, *Capital*. *Capital* has been the subject of two kinds of interpretation. For some—Schumpeter most recently —it is essentially a book of scientific economics without philosophical implications. For others, it is a kind of phenomenological or existential analysis of economics, and a few passages which lend themselves to a philosophical interpretation—for example, the chapter on commodity fetishism —supposedly provide the key to Marx's thought. Without entering into these controversies, I shall offer my own personal interpretation.

In my opinion, Marx regarded himself as a scientific economist. In the manner of the English economists on whom he was raised, he thought of himself as both heir to and critic of English political economy. He was convinced that he retained whatever is best in this economics, at the same time correcting its errors and transcending those limitations which may be imputed to the capitalist or bourgeois point of view. When Marx analyzed value, exchange, exploitation, surplus value, and profit, he wanted to be a pure economist, and he would not have dreamed of justifying some scientifically inaccurate or questionable statement by invoking a philosophical intent. Marx took science seriously, and I think we must do likewise.

But Marx was not an English economist of strict observance for a couple of very specific reasons which he has, in fact, indicated, and we need only recognize these in order to understand how to classify his work.

Marx reproached the classical economists for having considered the laws of capitalist economy universally valid. According to him, each economic regime has its own economic laws. The economic laws of the classical economists are, in the circumstances in which they are true, merely the laws of the capitalist regime. Hence the first important modification: we shift from the idea of an economic theory which is universally valid to the idea of the specificity of the economic laws of each regime.

Second, a given economic regime cannot be understood apart from its social structure. There are economic laws characteristic of each regime, because economic laws are, so to speak, the abstract expression of the social relations that define a certain mode of production. For example, in the capitalist regime, as we shall see, it is the social structure which explains the essential economic phenomenon of exploitation, and similarly it is the social structure which determines the inevitable self-destruction of the capitalist regime.

In other words, *Capital* represents an impressive attempt —and an attempt of genius, in the strict sense of the word —to account simultaneously for the mode of functioning of capitalism, the social structure of the capitalist regime, and the history of capitalism. Marx was an economist who wanted to be a sociologist at the same time: beginning with the understanding of how capitalism functions, we see why men are exploited in the regime of private ownership and why this regime is doomed by its contradictions to evolve toward a revolution which will destroy it.

The analysis of the functioning and evolution of capitalism also provides a sort of history of mankind in terms of modes of production. *Capital* is a book of economics which is at the same time a sociology of capitalism and also a philosophical history of man, caught up in his own conflicts until the end of prehistory.

I have said that this attempt is impressive, but I hasten to add that I do not think it is successful. No attempt of this kind has succeeded to date. By this I mean that today's economic or sociological knowledge makes use of valid partial analyses of capitalism's mode of functioning, of valid sociological analyses of the destiny of men or of classes in a capitalist regime, of certain historical analyses which account for the transformation of the capitalist regime—but there is no over-all theory uniting in a necessary way social structure, mode of functioning, the destiny of men in the regime, and the evolution of the regime. Why is there no theory that succeeds in encompassing the whole? Perhaps because this whole does not exist; that is, perhaps because history is not rational and "necessary" to this degree.

However this may be, to understand *Capital* is to understand how Marx sought to analyze simultaneously the functioning and the evolution of the regime and to describe the destiny of mankind within the regime.

As you know, *Capital* consists of three volumes. Only the first volume was published by Marx himself; Volumes II and III are posthumous. They were culled by Engels from Marx's many manuscripts and are far from being complete. The interpretations to be found in Volumes II and III are open to question because certain passages seem contradictory. I have no intention of summarizing here the whole of *Capital,* but it does not seem impossible to select the essential themes which are those Marx emphasized and are also those that have had the most influence in history.

The first theme, present from the beginning of *Capital,* is that the essence of capitalism is above all the pursuit of profit. Capitalism, insofar as it is based on private ownership of the means of production, is also based on the pursuit of profit by the entrepreneurs or producers.

When, in his last work, Stalin wrote that the fundamental law of capitalism was pursuit of maximum profit, while the fundamental law of socialism was satisfaction of the needs of the masses and the raising of their cultural level, he of course vulgarized Marx's theory from the level of higher education down to the level of elementary education, but he did retain the initial theme of Marxist analysis which is found in the first pages of *Capital*.[3] These pages contrast the two modes of exchange. There is one type of exchange which proceeds from commodity to commodity by way of money. You possess goods for which you have no use, and you exchange them for goods which you need, giving the goods you had to someone who wants them. This exchange may operate in a direct manner, in which case it is barter; or it may operate in an indirect manner, by way of money, which is the universal equivalent for merchandise.

The exchange which proceeds from commodity to commodity might be regarded as the immediately intelligible, immediately human exchange, but it is also the exchange which does not release any profit or surplus. As long as you proceed from commodity to commodity, you are in a relation of equality.

There is, however, a second type of exchange, one that proceeds from money to money by way of commodity and which has this peculiarity: that at the end of the process of exchange you have a greater sum of money than you had initially. Now, this type of exchange—proceeding from money to money by way of commodity—is the exchange characteristic of capitalism. In capitalism, the entrepreneur or producer does not proceed by way of money from a commodity he does not need to another commodity he does need; the essence of capitalist exchange is to proceed from money to money by way of commodity and to end up with more money than one had at the outset.

This type of exchange is in Marx's eyes the capitalist exchange par excellence and is also the most mysterious type of exchange. How is it that one can acquire something by exchange which one did not possess to begin with, or at least have more than one had to begin with? This leads to what is for Marx the central problem of capitalism, which might be stated as follows: Where does profit come from? How can there be a regime in which the essential impulse to activity is the pursuit of profit and in which producers and merchants are able, for the most part, to make a profit?

Marx is convinced that he has found a satisfactory answer to this question. By means of the theory of surplus value, he proves both that everything is exchanged at its value and that nevertheless there is a source of profit. The stages of this demonstration are these: theory of value, theory of wages, and finally surplus value.

First proposition: The value of any commodity is roughly proportional to the quantity of average human labor contained in it. This is what is called the theory of value.

Second proposition: The value of labor is measured in the same way as the value of any commodity. The wage the capitalist pays the wage earner in compensation for the labor power the latter rents to him is equal to the amount of human labor necessary to produce the merchandise indispensable for the existence of the worker and his family. Human labor is paid at its value, in conformity with the general law of value valid for all commodities.

Third proposition: The labor time necessary for the worker to produce a value equal to the one he receives in the form of wages is less than the actual duration of his work. Let us say that the worker produces in five hours a value equal to the one contained in his wage, and that he works ten hours. Thus he works half of his time for himself and the other half for the entrepreneur. Let us use the term "surplus value" to refer to the quantity of value produced by the worker beyond the necessary labor time, meaning by the latter the working time required to produce a value equal to the one he has received in the form of wages.

Thus we understand the origin of profit and how an economic system in which everything is exchanged at its value is at the same time capable of producing surplus value, i.e. —on the level of the entrepreneurs profit. There is a commodity which has the peculiarity of being paid at its value and at the same time of producing more than its value: namely, human labor.

It is easy to see how delighted Marx was with an analysis of this kind which seemed absolutely scientific, because (*a*) it explained profit in terms of an inevitable mechanism inherent in the capitalist regime, and (*b*) this same analysis lent itself to a denunciation and vituperation of capitalism, since it showed that the worker was exploited, that he worked part of his time for himself and the other part of his time for the capitalist or the entrepreneur. Marx was a scientist, but he was also a political prophet.

III

THE PROPOSITIONS that constitute the Marxist theory of exploitation may be summed up as follows:

(1) The value of a commodity is roughly proportionate

to the quantity of average human labor power crystallized in it.

(2) Labor power is rented at its value, and the value of labor power is determined by the value of those articles indispensable to the life of the worker and his family.

(3) The necessary quantity of work—i.e., the quantity of work needed to produce the value in the merchandise indispensable to the life of the worker and his family—is less than the total working day. As a consequence, the worker works part of the day for himself—that is, to produce a quantity of value equal to the one he receives in the form of wages—and another part of the day to produce the surplus value which will be appropriated by his employer.

(4) The part of the working day necessary to produce the value crystallized in his wage is called "necessary labor time"; the rest is called "surplus labor time." The value produced in surplus labor time is called "surplus value." And the rate of exploitation is defined by the relation between surplus value and the wages paid.

This theory of exploitation had a double virtue in Marx's eyes. First, according to him, it solves a problem inherent in the capitalist economy which may be formulated as follows: If there is equality of value in exchange, where does profit come from? Marx felt that, while solving a scientific enigma, he was also providing a logical basis for his protest against a certain kind of economic organization. Secondly, his theory of exploitation provided what we would call a sociological basis for the economic laws of the operation of the capitalist economy. Marx believed that economic laws are historical and that each economic regime has its own laws. His theory of exploitation is an example of such a historical law, because the mechanism of surplus value and exploitation presupposes the division of society into classes—of which one, that of the entrepreneurs or owners of the means of production, rents the labor power of the other, the workers—and the relation between these two classes is a *social* relation—a relation of power between two social categories.

How did Marx try to prove the propositions necessary to his theory of exploitation? The first proposition to be proved was that commodities are exchanged according to the quantity of average social labor crystallized in each. Marx did not claim that the law of value is exactly respected in every exchange; the price of a commodity fluctuates above and below its value according to the state of supply and demand. These fluctuations were not only not unknown to Marx; they were clearly stated by him. Further, Marx acknowledged that a commodity has value only to the extent that there is a demand for it. In other words, if a certain quantity of labor were crystallized in a commodity but no purchasing power were brought to bear on it, this commodity would cease to have value. Stated differently, the proportionality between value and quantity of labor presupposes, as it were, a normal demand for the commodity in question (and this, in effect, amounts to brushing aside one of the factors responsible for the fluctuations in price of the commodity).

Let us assume a normal demand for the commodity in question. Now, according to Marx, there is a certain proportionality between the value of this commodity, as expressed by its price, and the quantity of average social labor cyrstallized in the commodity.

Why is this the case? The main argument Marx gave is that labor is the only quantifiable element to be found in merchandise. If you consider use value, you are in the presence of a strictly qualitative element: there is no way to compare the usefulness of a fountain pen and the usefulness of a bicycle. Since we are seeking a basis for the exchange value of merchandise, we must find an element that is quantifiable, like the value itself. And the only quantifiable element, according to Marx, is the amount of labor contained, incorporated, or crystallized in each commodity.

Naturally there are difficulties which Marx himself recognized. Thus, the labor of the unskilled worker does not have the same value or the same creative potential of value as the labor of the foreman or the engineer. Admitting these

qualitative differences, Marx added that one need only reduce these different kinds of labor to a unity, which is average social labor.

The second proposition—namely, that the value of labor
is equal to the quantity of goods indispensable to the life of
the worker and his family—Marx gave as self-evident. (Ordinarily, when a proposition is given as self-evident, it is
because it is open to debate.) Marx says that since the
worker comes to the labor market in order to rent out his
labor power, the latter must be exchanged at its own value.
And, he says, this value must be measured in this case as
it is in all cases, i.e., by the quantity of labor needed to
produce it. Since Marx is interested in social rather than
biological science, he interprets human reproduction as
equivalent to human survival. Thus the quantity of labor
which will measure the value of labor power is that of the
commodities the worker and his family need to survive.

The trouble with this proposition is that, whereas the theory of labor value is based on the quantifiable character of
labor as a basis of value, in the case of the goods necessary
to the life of the worker and his family, we leave the realm
of the quantifiable. The necessary minimum will vary from
person to person, time to time, social circumstance to social
circumstance. This has led Schumpeter to declare that the
second proposition of the theory of exploitation is merely
a play on words.

If we accept the first two propositions, the third follows
on one condition: that the labor time necessary to produce
the value embodied in the wage be lower than the total
labor time; that is, that there be a discrepancy between the
working day and necessary labor time. Marx took this discrepancy between the working day and necessary labor
time for granted. He was convinced that the working day
in his time, which was ten and sometimes twelve hours, was
manifestly higher than the labor time necessary to create
the value embodied in the wage itself.

From this, Marx developed a casuistry of the struggle
over labor time. There are two fundamental methods of

increasing the rate of exploitation: one consists in increasing labor time, which, in the Marxists' schema, results in greater surplus labor time; the other consists in reducing necessary labor time to a minimum. One of the ways of reducing necessary labor time is by increased productivity, that is, by producing a value equal to that of the wage in fewer hours. Hence the mechanism that accounts for the tendency of a capitalist economy constantly to increase the productivity of labor; for an increase in the productivity of labor automatically reduces necessary labor time and, therefore, assuming the continuation of the level of nominal wages, increases the rate of surplus value.

Up to now I have discussed only the first volume of *Capital,* which is, you will remember, the only one published during Marx's lifetime. The two subsequent volumes consist of Marx's manuscripts, as published by Engels.

The subject of Volume II is the circulation of capital; it was to have explained how the capitalist economic system operated as a whole. In modern terminology we might say that, beginning with this microeconomic analysis of the structure and operation of capitalism in Volume II, Marx would have elaborated a macroeconomic theory comparable to Quesnay's *Tableau Économique,* with the addition of a theory of crisis. As a matter of fact, there are, scattered throughout Volume II of *Capital,* many elements of a theory of crises. But these elements do not themselves add up to a theory. It is possible, on the basis of the scattered indications in the second volume, to reconstruct and attribute various such theories to Marx. The only idea beyond question is that, according to Marx, the competitive, anarchic character of the capitalist mechanism and the necessity for the circulation of capital create a permanent possibility of disproportion between production and purchasing power. This is tantamount to saying that, essentially, an anarchic economy is characterized by crises. Are these crises regular or irregular? What is the combination of economic circumstances in which a crisis breaks out? On all these points, Marx gives hints rather than a precise theory.[4]

The third volume of *Capital* is basically the outline for a theory of the evolution of the capitalist regime, beginning with an analysis of its structure and operation. The central problem of the third volume is as follows. According to the plan of the first volume of *Capital,* in a given enterprise or a given sector of the economy, the more labor there is, the higher the surplus value; or again defining labor as "variable capital," the higher the ratio of variable to total capital, the higher the surplus value. In the schematism of the first volume, "constant capital"—the machines or the raw material—is transferred into the value of the goods without creating surplus value. All the surplus value proceeds from variable capital, or the capital that corresponds to the payment of wages. (The relation of variable capital to constant capital is called "the organic composition of capital." The relation of surplus value to variable capital is called "the rate of exploitation.") From this analytic relationship, one must conclude that in a given enterprise or sector, the more variable capital there is, the more surplus value there will be. In other words, there should be less and less surplus value as mechanization increases. Only it is obvious that this is not the case.

Marx was perfectly aware of the fact that appearances in the economy seem to contradict the fundamental relations he laid down in his schematic analysis. Until the third volume of *Capital* was published, Marxists and their critics both grappled with the question: If the Marxist theory of exploitation is correct, why is it that the enterprises and sectors with the highest ratio of constant to variable capital make the most profit? In other words, the apparent mode of profit seems to contradict the essential mode of surplus value.

Marx's answer is that the rate of profit is calculated, not in relation to variable capital, as is the rate of exploitation, but in relation to capital as a whole, i.e., the sum of constant capital and variable capital.

Why is it that the rate of profit is proportional not to surplus value but to the sum of constant and variable capital? Obviously, capitalism could not function if the rate of profit were proportional to variable capital. Indeed, there would be an extreme irregularity in the rate of profit, because the organic composition of capital, i.e., the relation of variable capital to constant capital, differs greatly from one sector of the economy to another. Thus, since the capitalist regime could not function otherwise, the rate of profit is actually proportional to capital as a whole and not to variable capital.

But why is it that the appearance of the mode of profit differs from the essential reality of the mode of surplus value? There are two answers to this question: the answer of the non-Marxists or anti-Marxists, and the official answer of Marx.

The answer of an economist like Schumpeter is simple: the theory of surplus value is false. That the appearance of profit is in direct contradiction to the essence of surplus value proves that the schematism of surplus value does not correspond to reality. When one begins with a theory and then discovers that reality contradicts this theory, one can, of course, reconcile the theory with reality by introducing a certain number of supplementary hypotheses; but there is another, more logical solution, which consists in recognizing that the theoretical schematism was badly constructed.

Marx's answer was that since capitalism could not function if the rate of profit were proportional to surplus value, an average rate of profit is substituted in each economy. This average rate of profit is a result of the competition between the enterprises and sectors of the economy. Competition forces profit to tend toward an average rate: there is no proportionality of rate of profit to surplus value in each enterprise or in each sector; but the total surplus value constitutes, for the economy as a whole, a sort of grand sum which is distributed among the sectors in proportion to the total capital, constant and variable, invested in each sector.

Why is this true? Because it cannot be otherwise; because if there were too great a disparity between rates of profit in different sectors, the system would not function. If there were a rate of profit of 30 percent or 40 percent in one sector and a rate of profit of 3 percent or 4 percent in another, capital could not be found to invest in the sectors where the rate of profit was low. Therefore, there must be established, through competition, an average rate of profit so that in the end the total surplus value is distributed among the sectors according to the amount of capital invested in each.

This theory of profit provides us with one of the main propositions of Marxist economics and also leads to the theory of evolution—what Marx called the law of "the falling tendency of the rate of profit."

Marx's point of departure was an observation which all the economists of his day made (or thought they made), according to which there was a perennial tendency toward a decline in the rate of profit. Marx, always eager to show the English economists how superior he was to them, thanks to his method, believed that in his schematism he had discovered the explanation for the historical phenomenon of the falling tendency of the rate of profit.[5]

Now, going back to elementary propositions: Average profit is proportional to capital as a whole, i.e., to the sum of constant capital and variable capital. But we also know that surplus value is deducted only from variable capital, i.e., human labor. We know further that the organic composition of capital changes with capitalist evolution and the mechanization of production and that the proportion of variable capital to total capital tends to diminish. This leads Marx to conclude that the rate of profit tends to decline proportionately as the organic composition of capital is altered, i.e., as the proportion of variable capital to total capital is reduced.

This law of the falling tendency of the rate of profit gave Marx a certain purely intellectual satisfaction, for he believed that he had demonstrated in a scientifically satisfactory manner a fact noted by observers but never, or badly, explained. In addition, he believed he had rediscovered what his master Hegel would have called "the cunning of Reason," that is, the self-destruction of capitalism by an inexorable mechanism functioning both through and beyond human influence. The competitive mechanism of an economy based on profit leads to the accumulation of capital, the mechanization of production, the reduction of the proportion of variable capital to total capital; and this in turn leads to a fall in the rate of profit, which in its turn leads to the doom of capitalism. You see that once again we encounter the fundamental pattern of Marxist thought: historical necessity acting through the influence of men but at the same time transcending the influence of each man —a historical mechanism leading to the destruction of the regime because of the intrinsic laws of its operation.

In my opinion, the center and the originality of Marxist thought lies precisely in this avowal of a necessity which is, in a sense, human but at the same time transcends all individuals. Each man, acting rationally in his own interest, contributes to the destruction of the interest common to all (at least common to all those who are interested in safeguarding the regime). The proposition is a sort of inversion of the essential propositions of the liberal economists. In the liberals' ideal representation of the economic world, each man, working in his own interest, works in the interest of the group. For Marx, each man, working in his own interest, contributes both to the necessary functioning and to the final destruction of the regime. The myth is still that of *The Communist Manifesto,* that of *The Sorcerer's Apprentice.*

There are several further ideas in *Capital* which I must mention before terminating this obviously brief and elementary analysis of the fundamental themes of the book and the problems it raises. Have we demonstrated yet that capitalism is self-destructive? What we *have* demonstrated is that the rate of profit tends to decline as a result of the modification of the organic composition of capital. But at what rate of profit is capitalism no longer capable of functioning? Strictly speaking, there is no answer to this in *Capital,* for no rational, schematic theory enables us to determine the specific rate of profit indispensable to the functioning of a particular regime.[6] In other words, strictly speaking, the law of the falling tendency of the rate of profit implies that the functioning of capitalism must become increasingly difficult, in proportion to mechanization or the increase of productivity. It does not demonstrate the inevitability of the final catastrophe; still less does it specify the moment it will occur. What, then, are the propositions which demonstrate the inevitable self-destruction of the capitalist system? Curiously, the only propositions which approach such a demonstration are the very ones to be found in *The Communist Manifesto* and in the works Marx wrote *before* he had made any detailed studies of political economy.

These propositions are those dealing with proletarianization and pauperization. Proletarianization means that, along with the development of the capitalist regime, the intermediate strata between capitalist and proletarians will be worn thin and that an increasing number of the representatives of these intermediate strata will be absorbed by the proletariat. Pauperization is the process by which the proletarians tend to grow poorer and poorer as the forces of production are developed. If we assume that, as more is produced, the purchasing power of the working masses is increasingly limited, it is indeed probable that the masses will have a tendency to rebel. According to this hypothesis, the mechanism of the self-destruction of capitalism is a sociological one and operates through the behavior of social groups.

There is also an alternative Marxist hypothesis: The income distributed to the masses is inadequate to absorb the increasing production, and there results a paralysis of the system, for the latter would be incapable of establishing an equilibrium between the commodities produced and the commodities demanded on the market by consumers.

There are, then, two possible representations of the capitalist dialectic of self-destruction: an economic dialectic, which is a new version of the contradiction between the constantly increasing forces of production and the relations of production that determine the income distributed to the masses, and a sociological dialectic functioning via the growing dissatisfaction and revolt of the proletarianized workers.

One question remains: How can pauperization be demonstrated, i.e., why in Marx's schematization must the income distributed to the workers diminish, absolutely or relatively, in proportion as the productive forces increase? It is not easy, even in Marx's scheme, to demonstrate pauperization. Now, according to *Capital,* the wage is equal to the quantity of commodities necessary to the life of the worker and his family. However, Marx hastened to add that what is necessary to the life of the worker and his family is not a matter of mathematically exact evaluation, but the result of a social evaluation which may change from society to society. But if we accept this social evaluation of the minimum standard of living, we must conclude that the workers' standard of living should rise. Indeed, it is likely that each society considers the minimum standard of living to be the one that corresponds to the productive possibilities of the society in question. Moreover, this is actually the case; the standard of living considered minimal in modern France or in the United States is considerably higher than it was a century ago.

Further, according to Marx himself, there is a way to raise the workers' standard of living without modifying the rate of exploitation. The increase in productivity need only permit the creation of a value equal to the wage in reduced necessary labor time. In the Marxist schema, productivity permits raising the workers' real standard of living without diminishing the rate of exploitation.

Thus, given the increase of productivity and the consequent reduction of necessary labor time, you end by excluding the rise of the real standard of living only by a theoretical augmentation of the rate of exploitation. But according to Marx the rate of exploitation is not raised! According to Marx, the rate of exploitation, at different periods, is nearly constant.

In other words, if one follows the economic mechanism as Marx has analyzed it, there is no proof of proletarianization, and one should rather draw one's conclusions from experience, i.e., the rise of the workers' real standard of living as a result of the increase in productivity and the modification of the social evaluation of a minimum standard of living.

Whence did Marx derive his proof of pauperization? The only proof, in my opinion, operates via a social mechanism, that of the so-called industrial reserve army. In Marxism, what prevents wages from rising is that there is a permanent surplus of unemployed manpower that weighs on the labor market and that modifies relations of exchange between capitalists and wage earners to the detriment of the workers.

In the theory of *Capital,* pauperization is not a strictly economic mechanism; it is an economic and sociological theory. Marx was not satisfied with the idea, common in his time, that, as soon as wages begin to rise, the birth rate increases and there also appears a surplus of workers on the labor market. There is also a second mechanism which is properly economic: The permanent mechanization of production tends to free a portion of the employed workers and as a result to create a kind of unemployment which today we call technological, that is, to create a reserve army which is the very expression of the mechanism whereby technologic and economic progress is made in capitalism. But, in that event, it is the sociological existence of the industrial reserve army that holds down the level of wages. Otherwise it would be possible to incorporate the historical fact of the workers' higher standard of living into the Marxist scheme without relinquishing the essential elements of the theory—although it would remove the element of inevitability.

It was undoubtedly one of Marx's ambitions to demonstrate that the destruction of capitalism was inevitable. My feeling is that, in *Capital,* reasons are given why the functioning of the system is difficult or, more accurately, why the functioning of the system becomes increasingly difficult, although this last proposition seems to me historically inaccurate. But it does seem to me that no proof of the self-destruction of capitalism is offered, unless it be via the revolt of the masses rebelling against their lot. If their lot should not, in fact, arouse extreme indignation, then *Capital* gives us no reason to believe that the destruction of the regime is in principle inevitable.

I hasten to add that all known economic and social systems were theoretically capable of surviving; nevertheless, they have disappeared. One ought not draw any premature conclusions from the fact that the death of capitalism was not theoretically demonstrated by Marx, for regimes have a way of vanishing without having been condemned to death by theorists.

IV

WHY DOES Marx's historical sociology of capitalism permit so many different interpretations? Why is it so ambiguous? Leaving aside accidental, historical, posthumous reasons—among them the destinies of movements and societies which have called themselves Marxist—the essential reasons for this ambiguity seem to me twofold. For one thing, the Marxist conception of capitalist society, and of society in general, is sociological, but this sociology is related to a philosophy; and a number of interpretative difficulties arise from the relation of a philosophy to a sociology. In addition, according to Marx, it is in terms of economic knowledge that a society as a whole is understood; but the relations between economics and sociology, or between economic phenomena and the social entity, are also ambiguous.

Let us start with a proposition that seems to me incontestable, or at least made obvious by all the texts. Marx came to political economy from philosophy by way of sociology, and until the end of his life he remained in a certain sense a philosopher. He always considered that the history of mankind, as it unfolds through the succession of regimes and culminates in a nonantagonistic society, had philosophical significance. It is through history that man creates himself. The culmination of history is at the same time a goal of philosophy. Through history, philosophy—by defining man—fulfills itself. The nonantagonistic, postcapitalist regime is not merely one social type among others; it is the goal, so to speak, of mankind's search for itself.

You will remember that I began my account of Marx with the mature works, or at least with those works written since 1847–1848. But there is a Marxian canon previous to this date, and I must now say a few words about the relation of Marxist thought to its philosophical origins.

Marx's thought is traditionally explained in terms of the conjunction of three influences, and it was Engels himself who named these three influences as decisive: German philosophy, English economics, and French history. This list of influences seems banal enough and is therefore scorned today by the more subtle interpreters. Let us begin with interpretations which are not subtle, that is, with what Marx and Engels themselves said about the origins of their thought.

According to them, they were in the tradition of classic German philosophy, retaining one of the main ideas of Hegelian thought, namely, that the succession of societies and regimes also represents the stages of philosophy and the stages of mankind.

Moreover, Marx studied the English economy; he availed himself of the ideas of the English economists; he adopted some of the accepted theories of his day: for example, the labor theory of value or the law of the falling tendency of the rate of profit. He believed that by adopting the concepts and theories of the English economists he would give a scientifically accurate formulation of capitalist economy.

As for the French historians, from them he borrowed the notion of the class struggle, which in fact was to be found almost everywhere in the historical writings of the end of the eighteenth century; but Marx, according to his own testimony, added a new idea, namely, that the division of society into classes is not a phenomenon associated with the whole of history or the essence of society but one that corresponds to a given phase. In a subsequent phase, the division into classes will disappear.[7]

These three influences certainly had their effect on Marx's thinking, and they provide a valid, if oversimplified, interpretation of the synthesis achieved by Marx and Engels. But this analysis of influences undoubtedly leaves most of the vital questions unanswered—especially the question of the relation between Hegel and Marx.

I shall not deal extensively with it here, for I should have to devote several hours to it and to presuppose the reader's familiarity with the works of Hegel, which would be unwise. I shall confine myself to a limited number of rather superficial remarks on the matter.

The first difficulty arises primarily from the fact that the interpretation of Hegel is at least as controversial as that of Marx. One may relate or contrast the two doctrines, depending on one's interpretation of Hegel's thought.

There is an easy way to produce a Hegelian Marx—which is to present a Marxist Hegel. This method is employed with a skill bordering on genius in Alexandre Kojève's book *Introduction à la Lecture de Hegel*. Here Hegel is Marxianized to such an extent that Marx's fidelity to Hegel's work can no longer be doubted.[8]

On the other hand, when someone, like my colleague M. Gurvitch,[9] does not like Hegel, he need only present him in the manner of the manuals of the history of philosophy —as an idealist philosopher who conceives of historical evolution as the evolution of the mind, as a succession of ideas very much removed from concrete phenomena—for Marx to become at once essentially anti-Hegelian. Gurvitch, in an attempt to reduce Marx's Hegelian heritage to a minimum, has given an interpretation of the origins of Marxist thought which is original and which places the emphasis on Marx's Saint-Simonianism. One chapter of Gurvitch's treatise is devoted to a demonstration (in my opinion a convincing one) of the Saint-Simonian influences on the thinking of the young Marx.

However, I am in disagreement with my colleague on one point, unfortunately the essential one. I do not doubt for a moment that Marx could have encountered Saint-Simonian ideas in his milieu, for the simple reason that these ideas, which we have already encountered in our study of Auguste Comte, were current in the Europe of Marx's youth and were to be found in one form or another almost everywhere, even in the newspapers. Since the early nineteenth century, sociological ideas have been in circulation, easily adopted by anyone interested in the movement of history. Marx was familiar with Saint-Simonian ideas, but he could not have borrowed from them what, in my opinion, is the heart of his own sociology.

What did he find in Saint-Simonianism? The opposition between two types of societies, the military and the industrial; the application of science to industry; the renovation of methods of production; the transformation of the world through industry. But the center of Marxist thought is not a Saint-Simonian or Comtist conception of industrial society; the center of Marxist thought is the contradictory character of capitalist industrial society. The idea of the intrinsic contradictions in capitalism is not contained in the Saint-Simonian or Comtist heritage. Neither Saint-Simon nor Auguste Comte believed that social conflict is the fundamental impulse of historical movement; neither believed that the society of his time was torn by insoluble contradictions. Because I feel the center of Marxist thought to be the contradictory character of capitalist society and the essential character of the class struggle, I refuse to regard Saint-

Simonianism as one of the major influences in shaping Marxist thought.[10]

But what does need clarification, I think, is how Marx believed he explained (1) that capitalist society was essentially contradictory and antagonistic, (2) where the contradictions came from, and (3) that the movement of history tended of itself to resolve this antagonism.

With this in mind, what are the Hegelian themes to be encountered in Marx's thought, in the youthful works as well as in those of his maturity?

The first fundamental idea—expressed in one of his theses on Feuerbach—is that philosophy is complete, and nothing remains but to realize it. Or again, that the only contribution still to be made to philosophy is to transcend it by realizing it. Or again, another proposition not equivalent but related to the foregoing: up to now, philosophers have conceived the world; the time has come to transform it.

What do these propositions mean in ordinary language? They mean that classical philosophy, culminating in Hegel's system, has reached an end; it is impossible to go further, because Hegel has conceived all of history and all of humanity. Philosophy has completed its task, which is to bring the experiences of humanity to explicit awareness. This awareness of the experiences of humanity is formulated in Hegel's *Phenomenology of Mind* as in his *Encyclopedia.*[11] But man, having become fully conscious of his past experiences and hence of his vocation, has not yet realized that vocation. Philosophy is complete as regards awareness, but the real world is not consistent with the meaning that philosophy gives to man's existence. Which raises what is, as it were, the original philosophical and historical problem of Marxist thought: Under what conditions can the course of history realize man's vocation as classical philosophy— i.e., Hegel's philosophy—has conceived it?

What is incontestably Marx's philosophical heritage is the conviction that historical movement has a fundamental meaning. A new economic and social regime is not just a new event presenting itself after the fact to the detached curiosity of professional historians; a new economic and social regime is a stage in the evolution of humanity itself.

If this is the central philosophical question, another question immediately arises: What is this human nature, this human vocation which history must realize in order for philosophy to realize itself?

To this question Marx's youthful writings offer various answers, all of which turn on a few positive concepts—universal man, total man—or, on the other hand, "alienation," a negative concept.

In the *Introduction to the Critique of Hegel's Philosophy of Law,* Marx tried to demonstrate that the universalization of the individual in accordance with the demands of Hegelian philosophy has not been achieved in the societies of his time. For—and again I shall express myself in a greatly oversimplified style—the individual, as he appears in Hegel's *Philosophy of Law*[12] and in the societies of his time, has a double position which is contradictory. On the one hand, he is a citizen. As a citizen, he participates in the state, i.e., in universality. But he is only a citizen once every four or five years, in the empyrean of formal democracy, as Marx put it, and he exhausts his citizenship—he completely fulfills his universality—in the vote. But outside of this single activity in which he fulfills his universality, he belongs to what Marx called the *bürgerliche Gesellschaft,* civil society, i.e., mainly professional activities, work, the infrastructure of society. Now, as a member of civil society, he is imprisoned in his particularities; he does not relate to the community as a whole. He is a worker in the service of an entrepreneur, or he is an entrepreneur separated from the collectivity. Civil society imprisons all individuals in their particularities and consequently prevents them from realizing their vocation of universality.

Under what condition will this contradiction be overcome? The answer is simple. Individuals must be able in their work itself to participate in universality in the same way they do in their activities as citizens, that is, as electors.

What do these abstract formulas mean? Formal democracy—to adopt the expression which has become popular in Marxist literature—is defined by the election of representatives of the people by all the citizens, and by abstract freedoms like freedom to vote or the freedom of discussion. But this formal democracy does not affect the working and living conditions of the members of the collectivity as a whole. The worker, who has only his wage to live on, who places his labor power on the market in exchange for a wage, bears no resemblance to the citizen who every four or five years elects his representatives and, directly or indirectly, his leaders. In order to achieve true democracy, the freedoms limited in modern societies to the political order would have to be transposed to men's concrete, economic existence.

Thus, individuals at work would have to be able to participate in universality as citizens do by means of the voting ticket. How could this real democracy be realized? Seemingly by abolishing the private ownership of the instruments of production, which places the individual in the service of other individuals, which in turn entails the exploitation of the workers by the entrepreneurs and prevents the entrepreneurs themselves from working directly for the collectivity, since in the capitalist system the entrepreneur works to obtain a profit.

In other words, the preliminary analysis contained in the *Introduction to the Critique of Hegel's Philosophy of Law* turns on the opposition between the particular and the universal, civil society and the state, the slavery of the worker and the fictitious liberty of the elector or citizen.[13]

The second concept which, as I have indicated, was central to Marx's youthful thinking is that of total man, probably more ambiguous than that of "universalized" man since, at least according to the interpretation I have given, the notion of universalization of the individual is not very mysterious.

One might say that for Marx, total man is the one who would not be mutilated by the division of labor. The man of modern industrial society, in the eyes of Marx, is actually a specialized man. He has acquired a specific form as a result of a particular trade. He remains imprisoned for the greater part of his existence by this specialized activity, and hence he leaves unused a number of aptitudes and capacities which might be developed.

According to this line of argument, total man is non-specialized man. And there are several passages in Marx which suggest a polytechnical training in which all individuals would be prepared for the greatest possible number of trades; after this training, individuals would be free not to do the same thing from morning to night.

Marx's writings contain several idyllic passages describing a future society in which men would go fishing in the morning, go to the factory in the afternoon, and go home and read Plato in the evening. This is not a ridiculous picture. I have known people working in *kibbutzim* in Israel who did read Plato in the evening. But this is a very exceptional case, associated with circumstances which are not the usual ones.[14]

One of the possible meanings of total man is man who is not cut off from certain of his aptitudes by the exigencies of the division of labor. According to this line of thinking, the notion of total man is a protest against the conditions imposed on the individual by industrial society—a protest which is both meaningful and sympathetic. For the division of labor does indeed have the result of not allowing the majority of individuals to realize all their capabilities. But this somewhat romantic protest does not seem very consistent with the spirit of a scientific socialism. Except in the case of an extraordinarily wealthy society in which the problem of poverty has been solved once and for all, it is difficult to imagine how any society, capitalist or otherwise, could train all individuals for all trades, or how an industrial society in which individuals were not specialized could function.

Hence, another direction has been explored for a less romantic interpretation. Total man cannot be man who is capable of doing everything, but man who truly realizes his humanity, who performs those activities which define man.

In this sense, the notion of work is essential, and the central problem becomes this: Man is essentially a creature who works; if he works under inhuman conditions, he is dehumanized, because he ceases to perform the activity that, given the proper conditions, constitutes his humanity. And in Marx's youthful writings, especially in the *Economic and Philosophic Manuscripts,* there is in fact a critique of capitalist working conditions.[15] In capitalism, man is alienated, and this alienation must be overcome before man can fulfill himself, can realize himself.

I should explain, for those who know German, that Marx made use of three different terms which, in the translations, are often rendered by the same word, "alienation," although the German terms do not have exactly the same meaning. They are *Entäusserung, Veräusserung,* and *Entfremdung.* The one that most closely corresponds to "alienation" in French or English is the last, which means to become a stranger to oneself, *fremd* meaning "stranger." *Entfremdung* is the activity or process by which someone becomes a stranger to himself. The idea is that under certain circumstances or in certain societies, the conditions imposed on man are such that he becomes a stranger to himself; he no longer recognizes himself in his activity or in his productions.

This notion of alienation obviously derives from Hegelian philosophy, where it plays a central role. But Hegelian alienation is conceived on the philosophical or metaphysical level. In the Hegelian conception the spirit, *Geist,* is itself alienated in its works; it constructs intellectual and social edifices and projects itself, so to speak, outside itself. The history of the mind, the history of humanity, is the history of these successive alienations, at the end of which the mind will find itself once again in possession of the whole of its works and of its historical past and will be aware of possessing this whole. In Marxism, including the writings of the young Marx, the process of alienation, instead of being a process which is philosophically or metaphysically inevitable, becomes the expression of a sociological process by means of which men or societies construct collective organizations in which they become lost.[16]

Alienation, sociologically interpreted, is at once a historical, moral, and sociological critique of the present social order. In the capitalist regime, men are alienated; they are themselves lost in the collectivity; and the root of all alienation is economic alienation.

What does economic alienation mean? It seems to me that, again in ordinary language, there are two modalities which correspond approximately to two criticisms Marx leveled against the capitalist system. A first form of alienation may be imputed to private ownership of the means of production, and a second to the anarchy of the market.

The alienation imputable to private ownership of the instruments of production manifests itself in the fact that work—the essentially human activity which defines man's humanity, as it were—has lost its human characteristics, because for the wage earner it has become merely a means of obtaining the wage necessary to maintain his existence. Instead of work's being the expression of man himself, work has been degraded into an instrument, a means of livelihood.

The entrepreneurs themselves are in a sense alienated, because the commodities they produce do not answer needs which are truly experienced by others but are put on the market in order to procure a profit for the entrepreneur. Thus the entrepreneur becomes a slave to an unpredictable market which is at the mercy of the hazards of competition. Exploiting the wage earner, he is not thereby humanized in his work, since he himself is alienated in the interests of the anonymous mechanism of the market.

Whatever the precise interpretation assigned to this idea of economic alienation, it seems to me clear that Marx's critique of the economic reality of capitalism was originally a philosophical and moral critique before becoming a strictly sociological and economic analysis. And one can understand the difficult problem of interpretation that arises here. I have indicated that Marx's thought may be presented purely and simply as that of an economist and sociologist, because at the end of his life Marx sought to be a scientist, economist, and sociologist. But there is no doubt that in his youth he arrived at economic and sociological criticism by way of philosophical themes. These philosophical themes—the universalization of the individual, total man, alienation—underlie the sociological analysis of the mature works. To what extent is the sociological analysis of Marx's maturity merely the development of the philosophical intuitions of his youth—or, on the contrary, does it utterly replace these philosophical intuitions? This problem of interpretation is still unresolved.

However, it seems clear that Marx certainly kept the philosophical themes I have just outlined in the background, if not in the foreground, all his life. For him, the analysis of capitalist economy was the analysis of the alienation of individuals and collectivities losing control over their own existence in a system subject to autonomous laws. His critique of capitalist economy was at the same time a philosophical and moral critique of the situation imposed on man by capitalism. Moreover, for Marx, the analysis of the evolution of capitalism was undoubtedly the analysis of the evolution of man and human nature throughout history; he expected from postcapitalist society the fulfillment of philosophy.

This granted, what was this total man which the postcapitalist revolution was to achieve? This point is highly debatable, because in Marx there is a fundamental fluctuation between two somewhat contradictory themes. According to one theme, man fulfills his humanity in work, and it is the liberation of work which will mark the humanization of society. On the other hand, there is occasionally another conception according to which man is truly free only outside of work. In this second conception, man realizes his humanity only to the extent that his labor time has been sufficiently reduced so that he has the possibility of doing something besides working.[17]

Of course, it is possible to combine these two themes by saying that the complete humanization of society would presuppose that, first, the conditions imposed on man in his work were humanized and that, simultaneously, his labor time was sufficiently reduced so that he could read Plato in his leisure hours. Philosophically, however, there remains one question: What is the essential activity which defines man and which must flourish before society can permit the realization of philosophy? If there is no definition of the essentially human activity, we may have to go back to the conception of total man in its vaguest sense. Society must permit all men to realize all their aptitudes. This statement represents a good definition of the ideal of society, but it is not easily translated into a concrete and specific program. Moreover, it is difficult to ascribe the fact that all men do not fulfill all their aptitudes solely to private ownership of the instruments of production.

In other words, there seems to be a great disproportion between the human alienation imputable to private ownership of the instruments of production and the fulfillment of total man which is to follow the revolution. How are we to reconcile the critique of modern society with the hope of achieving total man by the mere substitution of one mode of ownership for another?

Rapid though this analysis has been, you may have glimpsed what accounts for both the greatness and the ambiguity of Marxist sociology. It is essentially a sociology; it seeks to be a philosophy.

V

WITHIN THE philosophical ideas which were at the origin of Marx's thought and which constituted the basis for his sociology, there remain a number of obscurities or ambiguities which account for the many interpretations his thought has provoked. These ideas were essentially as follows: The history of social regimes is also the history of humanity—the history of man's creation of himself. Philosophy has conceived the world and man; now the problem is to fulfill what the philosophers have conceived. Philosophy must become sociology, and sociology must become revolutionary. The nonrealization of humanity in the world is alienation. The source of this alienation is in the economy and, more specifically, in private ownership.

The first ambiguity of a philosophical order concerns the nature of historical law. Indeed, we have seen that Marx's interpretation of history presupposes a meaningful evolution of a supraindividual order. Forms and relations of production are dialectically linked; through the class struggle and the contradiction between the forms and relations of productions, capitalism destroys itself. But this general vision of history can be interpreted in two different ways.

One interpretation I shall call objectivist. This representation of the historical contradictions leading to the destruction of capitalism and to the advent of a nonantagonistic society corresponds to what are commonly referred to as the broad outlines of history. From the confusion of historical data, Marx selected the essential facts, what is most important in the historical evolution itself, without including the details of events in this vision. If we accept this interpretation, the destruction of capitalism and the advent of a nonantagonistic society would be facts known in advance: certain, but at the same time indeterminate as to date and modality. But this type of foresight, in which capitalism will be destroyed by its contradictions, though when or how is not known, is not satisfying to the mind. Foresight involving an undated, unspecified event does not mean very much, or, at any rate, a historical law of this kind bears no resemblance to the laws of the natural sciences.

This is one of the possible interpretations of Marx's thought. It is the interpretation which is more or less orthodox in the Soviet world today. The necessary destruction of capitalism and its replacement by a more progressive society, i.e., by Soviet society, is asserted, but at the same time it is recognized that the date of this inevitable event is not yet known and that the manner of this anticipated catastrophe is still undetermined. This indeterminacy in the field of political events presents great advantages. For instance, one can declare in all sincerity that coexistence is possible. It is not necessary for the Soviet regime to destroy the capitalist regime, since the capitalist regime, one way or another, will destroy itself.[18]

The other interpretation of Marxist historical law is the one known, especially on Paris' left bank, as dialectical—and dialectical not in an ordinary sense but in a subtle sense. In this interpretation, the Marxist vision of history is born of a sort of reciprocal action between the historical world and the subject or consciousness that conceives this world. There would also be reciprocal action between the different sectors of the historical reality. This double reciprocity of action between subject and object and between the sectors of the historical reality would make it possible to understand events as they occur, in their concrete form.

If you consider the books of Jean-Paul Sartre or Merleau-Ponty, you will see that they retain some of the essential ideas of Marxist thought: the alienation of man in and by the private economy; the predominant influence of the forces and relations of production. But all these concepts, this whole schema of interpretation, is not intended to reveal historical laws in the scientific sense of the term or even the broad outlines of evolution. It is a way of making man's position in the capitalist regime intelligible, of relating events to man's position in the capitalist economic regime without there being determinism in the strict sense of the word.

A dialectical vision of this kind—of which there are several versions among the French existentialists and in the entire Marxist school linked with György Lukács—is philosophically more satisfying; but it, too, has its difficulties.[19]

The essential one is to recover the two fundamental ideas of elementary Marxism, namely, the alienation of man in capitalism and the advent of a nonantagonistic society after capitalism's self-destruction. A dialectical interpretation of reciprocal action between subject and object and between sectors of reality does not necessarily lead to these two essential propositions. It leaves unanswered the question: How do we determine which interpretation is true? If every historical subject conceives history in terms of his situation, why is the interpretation of the Marxists or of the proletariat true? Which suggests, if I may say so, the following alternative: The objectivist vision which invokes the laws of history involves the essential difficulty of declaring an undated and unspecified event to be inevitable; the dialectical interpretation can assert neither the necessity for revolution nor the nonantagonistic character of postcapitalist society nor the all-embracing character of historical interpretation.

The second philosophical ambiguity is this: Marx's thinking purports to be scientific, and yet it seems to imply imperatives; it prescribes revolutionary action as the only legitimate consequence of historical analysis. Whence a second duality of interpretation which may be summarized in the formula, Kant or Hegel? Must Marxist thought be interpreted in the context of the Kantian dualism of fact and value, or scientific law and moral imperative, or in the context of the monism of the Hegelian tradition?

In the posthumous history of Marxism, there is a Kantian and a Hegelian school, the latter being larger than the former. The Kantian school of Marxism is represented by Franz Mehring, a German social democrat who has written a biography of Marx, and by the Austro-Marxist Max Adler, who is more Kantian than Hegelian, but Kantian in a very special way.[20] However, the majority of Marx's interpreters have chosen to remain in the tradition of monism.

The Kantians argue that one cannot proceed from fact to value, from a judgment of reality to a moral imperative; hence one cannot justify socialism by an interpretation of history as it occurs. Marx analyzed capitalism as it is;

to advocate socialism involves a decision of a spiritual order.

The opposing school of Marxism asserts that the subject who understands history is engaged in history itself. Socialism, or the nonantagonistic society, must necessarily emerge from the present antagonistic society; moreover, the interpreter of history is led by a necessary dialectic from the observation of what exists to the desire for a society of another type. Certain interpreters, like Lucien Goldmann, go further and declare that in history there is no such thing as detached, disengaged observation of reality. For them, the vision of total history is inseparable from what they call an *engagement,* a commitment. It is as a result of one's desire for socialism that one perceives the contradictory character of capitalism. It is impossible to dissociate the taking of a position concerning reality from the observation of reality itself. Not that this taking of a position is arbitrary; it is produced through the dialectic of object and subject. Each of us is a part of history; it is from the historical reality that each of us selects his frame of reference and the concepts of his interpretation. Interpretation is born of our contact with the object—an object which is not acknowledged passively, however, but which is simultaneously acknowledged and denied, the denial of the object being an expression of our desire for another human reality.[21]

In abstract terms, we might say that there are two tendencies here: a tendency to dissociate the scientifically valid interpretation of history from the decision to be a socialist and, on the contrary, a tendency to make the interpretation of history inseparable from political desires.

Perhaps you are wondering, "And Marx?" Marx, as a man, was both scientist and prophet, sociologist and revolutionary. If he had been asked whether these two attitudes are separable, I personally think he would have answered that in the abstract they are. In my opinion, he was too scientifically oriented to admit that his interpretation of capitalism was bound up with a moral decision. But he was so thoroughly convinced of the worthlessness of the capitalist regime that for him the analysis of reality led inevitably to the desire for a revolution.

Beyond these two alternatives of Kant or Hegel, there exists a compromise which has today become the official philosophy of the Soviet world: the dialectical objectivist philosophy as expounded by Engels in his *Anti-Dühring*.[22]

What is dialectical materialism, according to the current official interpretation? The essential ideas are these:

(1) A dialectical conception declares that the law of reality is the law of change. There is constant transformation in inorganic nature as well as in the human world. There is no eternal principle; human and moral conceptions change from one age to the next.

(2) The real world is also characterized, as it were, by a hierarchy in the species of being, a sort of qualitative progression from inorganic nature to the human world, and in the human world from the initial regimes of humanity to the regime which will mark the end of prehistory, i.e., the socialist regime.

(3) Natural and social change occurs in accordance with certain abstract laws, of which the principal ones are these: Beyond a certain point, quantitative changes become qualitative. The transformations do not occur imperceptibly, a little at a time, but at a given moment there is a violent, revolutionary shift. Finally, the changes seem to obey an intelligible law: the law of contradiction or, more precisely, the law of the negation of negation.

What is the negation of negation? Here is an example from Engels: If you negate A, you have minus A; if you multiply minus A by minus A, you get A^2, which is, apparently, the negation of negation.

An example of the negation of negation in the human world: The capitalist regime is the negation of the regime of feudal ownership, and public ownership under socialism will be the negation of negation, i.e., the negation of private ownership.

In other words, and to translate, one of the characteristics of cosmic as well as of human movement would be the fact that changes are in a relationship of contradiction to one another and that this contradiction takes the following form: At moment B, there would be a contradiction of what existed at moment A, and moment C would contradict what existed at moment B and would in a sense represent a return to the original state of moment A, but on a higher level.

Another possible example of the negation of negation: First the initial collective ownership of property in archaic societies; the whole of history is the negation of this collective ownership in undifferentiated societies; and socialism negates social classes and antagonisms to return to the collective ownership of archaic societies, but on a higher level.

These dialectical laws have not completely satisfied all Marx's interpreters. There has been much discussion as to whether Marx was in agreement with Engels' materialist philosophy. The main question is this: To what extent is the notion of dialectics applicable to organic or inorganic nature as well as to the human world?

In the concept of dialectics, there is first the idea of change and then the idea of the relativity of ideas or principles to circumstances. But there are also the two ideas of totality and meaning. To achieve a dialectical interpretation of history, it is necessary for all the elements of a society or an age to form a whole and for the transition from one of these totalities to another to be intelligible, to have meaning. These two requirements of totality and intelligibility of sequence seem to be linked to the human world. One understands that, in the historical world, societies constitute total units, because the different aspects of the collectivities are in fact related to one another. The different sectors of a social reality may be explained in terms of one element regarded as essential, for example, the forces and relations of production. But, in organic or, above all, in inorganic nature, can we find the equivalent of these totalities? Can we find the equivalent of meaningful sequences?

In fact, this dialectical philosophy of the material world is by no means indispensable either to an acceptance of the Marxist analysis of capitalism or to being a revolutionary. One may not be convinced that minus A times minus A equals A^2 is an example of dialectics and still be an excellent socialist. The connection between the dialectical philosophy of nature as expounded by Engels and the center of Marxist thought is neither apparent nor necessary. Historically, a certain orthodoxy may combine these different propositions; but logically and philosophically, the economic interpretation of history and the critique of capitalism in terms of the class struggle have nothing to do with the dialectics of nature.

More generally, to what extent does the Marxist philosophy of capitalism imply metaphysical materialism? Here, again, I should say that the connection does not seem to me either logically or philosophically necessary. But the fact is that a number of politically active Marxists have believed that in order to be a good revolutionary it was necessary to be a materialist in the philosophical sense of the word. Since this belief was generally held by men who were very competent in revolutionary—if not in philosophical—matters, they probably had good reasons for it. Lenin, in particular, wrote a book called *Materialism and Empiriocriticism,* in which he tried to prove that those Marxists who abandoned a materialist philosophy were also straying from the royal road to revolution.[23]

Logically, there is no doubt one may be a disciple of Marx in political economy without being a materialist in the metaphysical sense of the word. Atheism, on the other hand, is related to the essence of Marx's Marxism, although one may be a believer and a socialist (but not a faithful follower of Marxism-Leninism). Historically, however, a kind of synthesis has been established between a philosophy of the materialist type and a vision of history.

I shall now turn to the second group of these problems of interpretation: those related to sociology. To some extent, we shall encounter the same type of uncertainties that I emphasized in discussing the first group, but I should like to show how, even aside from the philosophical basis of Marxist sociology, this sociology contains several ambiguities.

The center of the discussion is this: Marx's conception of capitalism in particular and of history in general depends on a combination of concepts—forces of production, relations of production, class struggle, class consciousness, or, again, infrastructure and superstructure—which may be used to analyze a given society. Personally, if I want to analyze a society, whether it be Soviet or American, I often begin with the state of the economy, and even with the state of the forces of production, and then proceed to the relations of production and finally to social relations. The critical and methodological use of these concepts to analyze a modern society, or perhaps any historical society, is unquestionably legitimate.

But if one confines oneself to utilizing these concepts in the analysis of societies, one does not thereby arrive at a philosophy of history. One risks finding that, at the same degree of development, productive forces may correspond to different relations of production. Private ownership does not exclude a high development of productive forces, and conversely, collective ownership may coincide with an inferior development of productive forces. But Marxist philosophy of capitalism and of history does presuppose a sort of parallelism between the development of the productive forces, the transformation of the relations of production, the intensification of the class struggle, and the march toward revolution.

The dogmatic conception of Marxism implies that the decisive factor is the force of production, that the development of the latter marks the direction of human history, and that the different stages in the development of the forces of production correspond to fixed stages of the relations of production and the class struggle. But suppose the class struggle is reduced with the development of the forces of production in capitalism? Or, again, suppose there is collective ownership in an underdeveloped economy? At once the parallelism between movements, indispensable to the dogmatic philosophy of history, collapses.

The same problem may be presented in abstract and, so to speak, epistemological terms. Marx seeks to understand all societies. Societies can be understood only in terms of their infrastructure, which is apparently the state of the productive forces, scientific and technical knowledge, industry and organization of labor. This understanding of societies, and above all of modern societies, in terms of their economic organization is entirely legitimate. But to shift from the analysis of societies in terms of the forces of production to a determined interpretation, one must admit determined relations between the different aspects of reality, between infrastructure and superstructure, between forces of production and relations of production. Marxists have felt that it was indeed difficult to use too-precise terms like *determination* in dealing with the relationship between forces and relations of production or, again, between the latter and the state of social consciousness. Since expressions of causality or determination have seemed too rigid or, as we say, mechanistic rather than dialectical, the term *conditioning* was immediately substituted for determination. As a formula, it is certainly preferable; but unfortunately the notion of conditioning is exceedingly vague. In a society, any sector conditions the others, the very law of social reality being that the different sectors condition each other reciprocally.

In ordinary language, if we had a different political regime, we should probably have, in certain respects, a different economic organization. If we had a different economy, we should probably have a different regime. If we had a different conception of economy, we should follow a different colonial policy. If we had a different colonial policy, we should have a different economic organization. And so forth. At a given moment, in given societies, the different sectors mutually condition each other.

Determination is too rigid a term, but *conditioning* risks being too flexible, because everyone grants that the state of the forces of production conditions relations of production and that the state of relations of production conditions class relations, the teaching of sociology at Harvard or the Sorbonne, and so forth. All this is incontestable—so incontestable that the usefulness of the term *conditioning* remains dubious.

What we need is an intermediate formula between *determination* of the whole of society by the infrastructure (a refutable proposition) and *conditioning,* which does not mean much.

As usual in such cases, the miraculous solution is the "dialectical" one. Conditioning is regarded as dialectical, reciprocal, with everything having an effect on everything else. But this loses its grasp on Marx's essential idea, namely, the determination of the social entity. On this point, I believe, Marx's thinking is quite clear. He believed that a historical regime was defined by certain major characteristics: the state of productive forces, the mode of ownership, and the relations between the workers and the people who take for themselves the surplus value. The different social types he recognized in history are each characterized by a certain mode of relations between owners (of slaves, of land, of means of production) and workers (slaves, serfs, wage earners). Slavery is one social type, wage earning is another. In other words, Marx believed he could find the specific characteristics of a historical state in terms of certain characteristics which in his eyes were fundamental. From this point on, there may indeed be dialectical relations between the different sectors of reality; but what remains essential, for Marx, is the definition of a social regime in terms of a certain number of facts regarded as decisive. The trouble is that these different facts, which are in Marx's eyes decisive and interrelated, seem to be separable. History, in fact, has separated them. And while dialectics can put them together again, after one ingenious fashion or another, it can never reestablish the original Marxist unity.

The ambiguity of Marx's sociology may also be revealed by an analysis and discussion of its essential concepts. Let us take, for example, the two terms *infrastructure* and *superstructure*. What are the elements of social reality which belong to the infrastructure? What are the ones that pertain to the superstructure? In general, it seems that infrastructure should refer to the economy, particularly the forces of production. But what are these so-called forces of production? All the technical apparatus of a civilization is inseparable from scientific knowledge; and the latter, in turn, seems to belong to the realm of ideas, of knowledge, and these last elements should derive from the superstructure, at least to the extent that scientific knowledge is, in many societies, inseparable from the way of thinking, from philosophy and ideology.

In other words, there are already present in the infrastructure, defined as forces of production, elements which should derive from the superstructure. This fact in itself does not imply that one cannot analyze a society by considering in turn the infrastructure and the superstructure. But it is exceedingly difficult to separate what belongs, according to the definition, to each.

Further, the forces of production depend not only on the technical apparatus but also on the organization of collective labor. The organization of collective labor depends, in turn, on the laws of ownership; the laws of ownership belong to the legal domain; the law is part of the reality of the state (at least according to certain passages);[24] and the state or politics seems to belong to the superstructure. Once again, we perceive the difficulty of truly separating what is infrastructure from what is superstructure. The discussion of what belongs to one or the other of these two terms can go on indefinitely.

These two concepts, as simple instruments of analysis, may, like any concepts, have a legitimate use. The objection applies only to a dogmatic interpretation in which one of the two terms determines the other.

In this discussion of infrastructure and superstructure, I have already anticipated the analysis and possible discussion of the terms *forces of production* and *relations of production*. We all know that one of the dialectics that plays the greatest role in the thinking of Marx and the Marxists is precisely the possible contradiction between the forces of production and the relations of production. But each of these terms presents ambiguities. Moreover, it is not easy to state precisely what constitutes the contradiction between the forces and relations of production. One of the simplest versions of this dialectic would be the following.

At a certain degree of development of the forces of production, individual right of ownership represents an impediment to their progress. In this case, the contradiction is between the full expansion of the technique of production and the preservation of individual right of ownership.

This contradiction contains, it seems to me, a share of truth, but it is not relevant to the dogmatic Marxist interpretations. If you consider the great modern enterprises in France, such as Citroën, Renault, or Péchiney, or Dupont or General Motors in the United States, you can say, in effect, that the volume of the forces of production has made it impossible to maintain individual right of ownership. The Renault factories may be said to belong to no one, since they belong to the state (not that the state is no one, but the state's ownership is abstract and fictitious, as it were). One might also say that Péchiney belongs to no one, since Péchiney belongs to thousands of shareholders, and while the latter are owners in the legal sense of the word, they no longer exercise the traditional and individual right of ownership. In the same way, Dupont and General Motors belong to hundreds of thousands of shareholders, who maintain the legal fiction of ownership but do not exercise its true privileges.

There is a passage in *Capital* in which Marx referred

to the great associations of shareholders, observed that individual ownership is disappearing, and concluded that the modern corporation is already overcoming typical capitalism (but, he added, without breaking the cadres of the system).[25] It may be said, therefore, that Marx was right to show the contradiction between the development of the forces of production and individual right of ownership since, in the modern capitalism of the great shareholding associations, right of ownership, in a certain sense, has disappeared.

On the other hand, if one believes that the great modern corporations, the great shareholding associations, are the very essence of capitalism, then it is easy to demonstrate that the theoretical contradiction between forces and relations of production does not exist. The development of the forces of production requires the appearance of new forms of the relations of production or, again, new forms of the traditional right of ownership.

According to the second interpretation of the contradiction between the forces and the relations of production, the distribution of income determined by individual right of ownership is such that a capitalist society is incapable of absorbing its own production. In this case, the contradiction between forces and relations of production affects the very functioning of a capitalist economy. The purchasing power distributed to the masses would always remain lower than the demands of the economy itself. This version of the contradiction between forces and relations of production is one that continues to be in vogue. I shall confine myself here to a common-sense observation: This contradiction between forces and relations of production has been exposed for a century or a century and a half; meanwhile, the forces of production in all capitalist countries have undergone a prodigious development. In other words, the incapacity of an economy based on private ownership to absorb its own production was already predicted when the forces of production were a fifth or a tenth of what they are today; and this will probably still be the case when the forces of production are five or ten times what they are today—which would seem to indicate that the contradiction has not been clearly demonstrated. It continues to be unknown at what point an economy in which private ownership persists is incapable of absorbing its own production.

In other words, neither of the two versions of the contradiction between forces and relations of production has been demonstrated. The only version that obviously contains a share of truth is the one that does not lead to those political and messianic propositions on which the Marxists insist most strongly.

Marx's sociology, as we have seen, presents another aspect or, at any rate, lends itself to an interpretation which is complementary to the one I have just been examining. Marx's sociology is, in fact, a sociology of the class struggle. Certain propositions are central, fundamental: modern society is an antagonistic society; classes are the principal actors in the historical drama of capitalism in particular and of history in general; the class struggle is the moving power of history and leads to a revolution which will mark the end of prehistory and the appearance of a nonantagonistic society.

Marx himself wrote, toward the end of his life, that he had found the idea of class and class struggle in the bourgeois historians, especially in the French ones, but that his own contribution to the theory of classes consisted of the following three propositions: The existence of classes is connected only with certain historical phases in the development of production; the class struggle leads inevitably to the dictatorship of the proletariat; this dictatorship is, in turn, merely a transitional stage in the abolition of all classes, in the realization of the classless society.

Given these propositions, which are central to Marx's thought, it seems to me that I should raise the most basic question, the one I did not raise when I was presenting Marx's thought because it is so difficult a question to answer: What is a social class? A great many passages can be found in Marx on this point. There is a classic passage which everyone knows, because it occurs on the last page of the manuscript of *Capital*. (The final chapter of the third volume of *Capital* is entitled "Social Classes.") Since *Capital* is Marx's principal scientific book, we must consider this passage, which is, unfortunately, incomplete.

In it, Marx distinguished three classes, related to the three sources of income: owners of simple labor power, owners of capital, and landowners, whose respective sources of income are wages, profit, and ground rent. Hence salaried workers, capitalists, and landowners form the three great classes of modern society, based on the capitalist mode of production.[26]

This analysis of classes in terms of their economic structure is the one that best answers Marx's scientific intention. From it, we may elucidate one or two of Marx's essential ideas regarding classes.

(1) A social class is that which occupies a fixed place in the process of production. A place in the process of production has two meanings, moreover: a place in the technical process of production, and a place in the legal process superimposed upon the technical one. The capitalist, owner of the means of production, is at the same time master of the organization of labor, master in the technical process, and also, because of his legal position, the one who takes the surplus value from the associated producers.

(2) Class relationships tend to become simpler with the development of capitalism. Indeed, if there are only two sources of income, aside from ground rent whose importance diminishes with industrialization, there are only two large classes: the proletariat, consisting of those who possess only their labor power, and the capitalists, those who appropriate a portion of the surplus value.

The second category of Marx's texts relating to classes consists of his historical studies, most of which are admirable; for example, his study of the German Revolution of 1848 or his study of Louis Bonaparte's *coup d'état* of the

Eighteenth Brumaire. In his historical studies, Marx utilized the notion of class, but without making it into a systematic theory. The enumeration of classes in Germany or France is longer and more complete than the structural distinction between classes to which I have referred.[27]

For example, in *Revolution and Counter-Revolution in Germany,* Marx distinguished the following classes: the feudal nobility, the bourgeoisie, the *petite bourgeoisie,* the upper and middle peasantry, the free lower peasantry, the slave peasantry, the agricultural laborers, and the industrial workers. In *The Class Struggle in France,* the list is as follows: financial bourgeoisie, industrial bourgeoisie, *petite bourgeoisie,* peasant class, proletarian class, and finally what he calls the *Lumpenproletariat,* which more or less corresponds to what we call the subproletariat.

This enumeration does not contradict the theory of class outlined in the last chapter of *Capital.* The problem Marx raised in these two kinds of passages is not the same. In one case, he was trying to ascertain the large groupings characteristic of a capitalist economy; in others he was trying to ascertain the social groups that have exerted an influence on political events in particular historical circumstances.

It is true, nevertheless, that it is difficult to effect the transition from the structural theory of class, based on the distinction between sources of income, to the historical observation of social groups. In fact, a class does not constitute a unity simply because, from the point of view of economic analysis, its income has a single and identical source; from all appearances, there must also be a certain psychological community and possibly a certain sense of unity or even a desire for common action.

This observation brings us to a third category of Marxist texts; and here I shall quote a classic passage from *The Eighteenth Brumaire.* In this passage, Marx explains why a large group of men, even if they share the same economic activity and the same style of life, do not necessarily represent a social class. Here is the passage.

The peasants are an immense mass whose individual members live in identical conditions, without however entering into manifold relations with one another. Their method of production isolates them from one another, instead of drawing them into mutual intercourse. . . . In so far as millions of families live under economic conditions that separate their mode of life, their interests and their culture from those of the other classes, and that place them in a hostile attitude toward the latter, they constitute a class. In so far as there exists among these peasants only a local connection in which the individuality and exclusiveness of their interests prevent any unity of interest, national connections and political organization among them, they do not constitute a class.

The idea—a very important one—is that community of activity, way of thinking, and mode of life is a necessary but insufficient condition for the reality of a social class. For there to be a class, there must be a consciousness of unity, a feeling of separation from other social classes, and even a feeling of hostility toward other social classes. Which explains a shorter and more categorical passage: "Separate individuals form a class only to the extent that they must carry on a common struggle against another class."

If we take all these passages into consideration, it seems to me that we arrive, not at a complete and professorial theory of class, but at a political and sociological theory of class which is sufficiently clear.

Marx's original idea was a fundamental contradiction of interests between wage earners and capitalists. He was convinced, moreover, that this fundamental opposition of interests dominated all of capitalist society and would assume an increasingly simplified form in the course of historical development.

From another point of view, Marx, as an observer of historical reality (and he was an excellent one), was extraordinarily aware of the plurality of social groups, a plurality not reducible to two large groups, i.e., capitalist on the one hand and wage earners on the other. But class, in the true sense of the word, is not to be confused with any ordinary social group. Social class, in the true sense of the word, implies, beyond a community of existence, the consciousness of this community and the desire for common action with a view to a certain organization of the collectivity. And on this level, it is clear that in Marx's eyes there are in effect only two great classes, because there are, in capitalist society, only two groups which have truly contradictory images of what society should be and have also a definite political and historical purpose. These two groups are the wage earners and the capitalists.

In the case of the workers versus the owners of the means of production, the various criteria which may be invented or observed are identified. The industrial workers have a determined mode of existence which depends on the lot they are assigned in capitalist society. They are conscious of their solidarity; they are becoming conscious of their antagonism toward other social groups. They are, therefore, a social class in the true sense of the word, a class which is politically and historically defined by a will of its own. The proletariat's will places it in fundamental opposition to the capitalists. There are sub-groups within each of these classes and also groups which are not yet absorbed into the camp of one or the other of the two chief actors in the drama of history. But these exterior or marginal groups, the merchants, the *petite bourgeoisie,* the survivors of a former social structure, will gradually, in the course of historical evolution, be obliged to join one or the other of the two existing camps: the camp of the proletariat or the camp of capitalism.

These, I think, are Marx's major ideas regarding social classes. What in this theory is most open to debate or misunderstanding? It seems to me that two points, central to Marx's thought, are ambiguous and debatable.

One might say that the point of departure of Marxist analysis is the parallel between the rise of the bourgeoisie and the rise of the proletariat. In his early writings, Marx described the advent of a fourth estate as analogous or similar to the rise of the third. The bourgeoisie developed new forces of production within feudal society. In the same way, the proletariat is developing new forces of production within capitalist society. But this analogy seems to me to be false. One must have political passion, as well as genius, not to see that the two cases are radically different.

The bourgeoisie, whether commercial or industrial, which created new forces of production within feudal society was really a new social class formed within the old society. But the bourgeoisie, whether commercial or industrial, was a privileged minority which performed socially dominant functions. The bourgeoisie opposed the feudal ruling class as an economic aristocracy opposes a military aristocracy. Hence this socially unprecedented privileged class was able to create new forces and relations of production within feudal society; at a certain moment in history, this socially privileged class overthrew the political superstructure of feudalism. The French Revolution, in Marx's eyes, represented the moment when the bourgeois class seized the political power still in the hands of the remnants of the politically dominant feudal class.

Let us now consider the proletariat. In capitalist society, the proletariat is not a privileged minority; it is the great mass of unprivileged laborers. The proletariat does not establish new forces or relations of production within capitalist society. The workers, the proletarians, are the agents of execution of a system of production directed either by capitalists or by technologists.

Therefore, the analogy between the rise of the proletariat and the rise of the bourgeoisie is sociologically false. In order to restore the equivalence between the rise of the bourgeoisie and the rise of the proletariat, the Marxists are forced to resort to something which they themselves condemn when practiced by others, namely, myth. For in order to link the rise of the proletariat with the rise of the bourgeoisie, one must identify the minority ruling the political party in the name of the proletariat with the proletariat itself.

In other words, in the last analysis, in order to maintain the analogy between the rise of the bourgeoisie and the rise of the proletariat, Lenin, Stalin, and Khrushchev must each in turn represent the proletariat. In the case of the bourgeoisie, it is the bourgeois who are privileged, who control commerce and industry, who rule. But when the proletariat has its revolution, it is men claiming to represent the proletariat who control commerce and industry and who exercise power.

The bourgeoisie is a privileged minority which passed from a socially dominant position to the exercise of political power; the proletariat is the great unprivileged mass which cannot, as proletariat, become the privileged and dominant minority, though of course political parties or groups of men may claim to represent the proletariat in order to establish a new regime.

Do not misunderstand me; I make no value judgment as to the respective merits of a regime claiming to represent the bourgeoisie and one claiming to represent the proletariat. All I should like to establish here, because to me these are facts, is that the rise of the proletariat cannot, except by mythology, be compared with the rise of the bourgeoisie and that herein lies the central, immediately obvious error of the entire Marxist vision of history, an error whose consequences have been considerable.

This brings us to the difficulties inherent in the Marxist relation between sociology and economics. First of all, as I have tried to explain, Marx wanted to combine a theory of the functioning of the economy with a theory of the evolution of the capitalist economy. This synthesis of theory and history contains two intrinsic difficulties, one at the outset, and one at the end. The initial difficulty is this: The capitalist regime, as Marx described it, can function only if there exists a group of men who possess available capital and consequently are in a position to rent the labor power of those who possess nothing else. This raises a historical question: How does this group of men come to be? What is the formative process of the original accumulation of capital which is indispensable to the functioning of capitalism itself?

It is not difficult to explain historically the formation of this group of capitalists. Violence, force, guile, theft, and other procedures traditional in political history easily account for the formation of a group of capitalists. The difficulty is to explain in *economic* terms the formation of this group that is indispensable to the functioning of capitalism. In other words, an analysis of the functioning of capitalism presupposes at the outset extraeconomic phenomena in order to create the conditions under which the regime can function.

The same difficulty appears at the conclusion. If you will recall, I tried to explain Marx's notion of the mechanism of self-destruction inherent in capitalism, and I also showed how, in the last analysis, there was no conclusive demonstration either of the moment when capitalism will cease to function or even of the fact that at a given moment, it will cease to function. For the economic demonstration of the self-destruction inherent in capitalism to be conclusive, the economist should be able to say, with reference to the law of the falling tendency of the rate of profit, that capitalism cannot function at a rate of profit below a certain percentage; or again, that after a certain point, distribution of income is such that the regime is incapable of absorbing its own production. But, in fact, neither of these two demonstrations is to be found in *Capital;* even the doctrine of increasing pauperization is not demonstrated in the economic analyses of *Capital.*

In other words, Marx gave a certain number of reasons for believing that the capitalist regime would have to function badly, but there is no economic demonstration of the destruction of capitalism because of its internal contradictions; hence it becomes necessary, I think, to introduce at the end as well as at the beginning of the process a factor external to capitalism itself, which must be of a political order.

Secondly, there is an essential difficulty about the theory of capitalist economy as an economy of exploitation. The capitalist theory of exploitation is based on the notion of surplus value. The notion of surplus value is, in turn, inseparable from the theory of wages. Now, every modern economy is a progressive economy. This means that every modern economy must accumulate a part of the annual production with a view to expanding the forces of production; or, to use modern terminology, with a view to augmenting the machinery of production. Thus, if capitalist economy is defined as the economy of exploitation, the problem is to show how and to what extent the capitalist mechanism of saving and investment, or again the capitalist mechanism of accumulation, differs from the mechanism of accumulation which exists or would exist in a modern economy of another type.

Marx wrote, "Accumulate, accumulate, that is the law and the prophets." In his eyes, the characteristic of capitalist economy is a higher rate of accumulation of capital.[28] Yet let us consider an economy of the Soviet type. You know that one of the merits which theorists of Soviet economy claim for the latter is the high percentage of accumulation.

A century after Marx, the ideological competition between the two regimes is focused on the rate of accumulation practiced by each. All well and good. But the question is whether the capitalist mechanism of accumulation is better or worse than the mechanism of accumulation of another economic regime (better for whom, and worse for whom?).

In abstract terms, the problem is this: In his analysis of capitalism, Marx considered simultaneously the characteristics of all economies and the characteristics of a modern economy of the capitalist type, because he knew no other. A century later, the true problem of the true Marxist would be to analyze the peculiarities of a modern economy of the capitalist type in relation to the peculiarities of a modern economy of another type.

The theory of wages, the theory of surplus value, the theory of accumulation, cease to be entirely satisfactory in themselves. Rather, they represent questions raised, or analytical points of departure, in order to differentiate what might be called capitalist exploitation from Soviet exploitation or, to employ a more neutral terminology, to differentiate capitalist surplus value from surplus value in the Soviet regime. For, in any regime, it is impossible to give the workers all the value they produce because a part of this value must be set aside, on the one hand for the wages of management, and on the other hand for collective accumulation. There remain, naturally, important differences between the two mechanisms, since in the capitalist regime accumulation proceeds via individual profit and since the distribution of income is not the same in the two regimes.

VI

THE VARIOUS critical remarks I have directed toward Marxism, criticisms that are easy to make a century after Marx, imply no claim—which would be ridiculous—to superiority. I merely want to show that Marx, observing the beginnings of the capitalist regime, was not able to distinguish easily between what is implied by a regime of private ownership, what is implied by the phase of development of a modern economy which the English economy was going through at the time he was observing it, and finally what is implied by any industrial economy.

Today, the task of economic analysis, sociologically speaking, is precisely to discriminate among these three kinds of elements: characteristics of any modern economy, characteristics of a particular system of modern economy, and finally characteristics linked to one phase of growth of the modern economy. Such discrimination is difficult, since in reality all these characteristics are always present and combined. But if one seeks to make a political or moral judgment of a certain system, obviously one must not ascribe to the system that which is imputable either to the general characteristics of modern economy or to a specific phase of its development.

The very type of the confusion between these different elements is the theory of capital accumulation and surplus value. Every modern economy implies accumulation. The rate of accumulation is higher or lower according to the phase of development and also according to the intention of the government of the society in question. What does vary, on the other hand, is the economic and social mechanism of surplus value, i.e., the mode of circulation of savings. A planned economy may obtain a flow of savings of a determined type, while an economy in which private ownership of the instruments of production persists admits of a more complex mechanism and does not readily tolerate the authoritarian determination of the amount of savings and, thereby, of the percentage of accumulation in relation to total national income.

I come now to a final aspect of the relation between the economic and the sociological analysis, namely, the problem of the relation between the political regime and the economic system. In my opinion, it is on this point that Marx's sociology is weakest.

For what do we find on this decisive problem in *Capital,* as well as in Marx's other works? A few familiar ideas, always the same ones. The state is considered essentially as the instrument of domination of one class, the instrument by which one class exploits the others. In contrast to the economic and social regime, consisting of antagonistic classes and of the domination of one class over the others, Marx paints the picture of an economic and social regime in which there will no longer be class domination, or political power exercised by one class over the others. After the antagonistic society, there will appear a nonantagonistic society and, as a consequence (by definition, as it were), the state will have to disappear, since the state exists only

212

so long as one class needs to dominate and exploit the others.

Between the antagonistic society and the nonantagonistic society of the future is interposed what is called the dictatorship of the proletariat, a phrase which occurs twice in Marx's works, particularly in a famous text, *The Critique of the Gotha Program* (a program established by the German socialist party).[29] The dictatorship of the proletariat is defined as the final strengthening of the state before the crucial moment when the state itself must perish. Before disappearing, the state attains, so to speak, its culmination.

The dictatorship of the proletariat was not very clearly defined in Marx's writings, which offered two versions of the idea. One was that of the Jacobin tradition, namely, the absolute power of a party claiming to represent the masses. The other, almost opposite, version had been suggested to Marx by the experience of the Commune of Paris, which tended toward the decentralization of political authority.

This conception of politics and of the disappearance of the state in a nonantagonistic society seems to me by far the most easily refutable sociological conception in all of Marx's work. Without indulging in any sort of polemics, the reasons why it is easy to criticize this theory seem to me to be these.

(1) No one denies that in any society—and particularly in a modern society—there are common functions of administration and authority which must be performed. No one can reasonably suppose that an industrial society as complex as our own can do without an administration, and an administration which is centralized in certain respects.

Moreover, to the extent that we presuppose a planned economy, it is inconceivable that there not be centralized organisms to make the fundamental decisions implied by the very idea of planning. But these economic and social decisions made by central organisms of planning presuppose functions which are commonly called functions of state. Therefore, unless we imagine a period of absolute abundance in which the problem of the co-ordination of

213

production no longer arises, a regime of planned economy requires a reinforcement of the administrative and directorial functions performed by the central power.

In this sense, the two ideas of a planned economy and the disappearance of the state are contradictory for the foreseeable future, so long as it is important to produce as much as possible, to produce according to the directives of planning, and to distribute production among the social classes at the discretion of those in power.

Therefore, if the word *state* refers to all the administrative and directorial functions of the collectivity, the state cannot disappear in any industrial society, let alone a planned industrial society, since by definition central planning implies that a greater number of decisions are made by the government than in a capitalist economy, which is partially defined by the decentralization of the decision-making power.

(2) Hence, the disappearance of the state can have only a symbolic meaning. What *does* disappear is the class character of the state in question. It is, in fact, conceivable that from the moment class rivalry ceases to exist, these administrative and directorial functions, instead of expressing the dominant intention of a particular group, become the expression of the society as a whole. In this sense, one can in fact imagine the disappearance of class character, of domination and exploitation, from the state itself.

But beyond this political interpretation, a new question arises: Can the state, in the capitalist regime, be defined solely and essentially in terms of the power of a given class?

We have seen that Marx's central idea is that capitalist society is antagonistic. All the essential characteristics of this regime proceed from this antagonism. But the question arises as to whether or how there could be a society without antagonism. The whole argument rests on the qualitative difference between the bourgeois class, which exercises power when it possesses the instruments of production, and the proletariat, considered as the class which is to succeed the bourgeoisie.

Karl Marx

I have explained why, in my opinion, the comparison between the coming to power of the bourgeoisie and the coming to power of the proletariat does not hold true in sociological terms. As for the relation between the economy and politics in a nonantagonistic society, the same question arises again in the following form: To say that the proletariat is a world class which, at some point, assumes power can have only a symbolic significance, since the mass of factory workers cannot be confused with the dominant minority which exercises power. Consequently, the expression "the proletariat in power" is merely a symbolic way of referring to the party or group of men claiming to represent the masses.

As for the nonantagonistic society, the problem is that in a society in which there is no longer private ownership of the instruments of production, by definition there is no longer any antagonism connected with this ownership; but there are men who exercise power in the name of the masses. There is, therefore, a state which performs the administrative and directorial functions indispensable to any developed society. A society of this type is not characterized by the same antagonisms as a society in which there is private ownership of the instruments of production. But a society in which the state, by means of economic decisions, largely determines the condition of each and every man may obviously be characterized by antagonisms between groups, whether these be horizontal groups—peasants versus workers—or vertical groups—those at the bottom and those at the top of the hierarchy.

Understand, I am not saying that in a society in which economic conditions depend on planning and the planning is done by the state, conflict is *inevitable*. I am simply saying that one cannot establish the basis for a nonantagonistic society on the mere fact that private ownership of the instruments of production has disappeared and each man's condition depends on the decisions of the state. Because the decisions of the state are made by individuals or by a minority, these decisions may correspond to the interests of particular groups. There is no preestablished harmony between the interests of different groups in a planned society.

Thus, the power of the state does not and cannot disappear in a planned society, even when private ownership of the instruments of production has disappeared. A planned society can be governed in an equitable manner by the planners, but there is no guarantee that the latter will make decisions which correspond either to the interests of all or to the highest interests of the collectivity, insofar as these can be determined.

The guarantee of the disappearance of antagonisms would presuppose either that intergroup antagonisms have no other basis than private ownership of the instruments of production or that the state disappears. But each of these two hypotheses is, to say the least, unlikely. There is no reason why all the interests of the members of a collectivity should become harmonious simply because the instruments of production cease to be private property. One type of antagonism disappears, but not all possible antagonisms. Furthermore, as long as administrative and directorial functions persist, there is by definition the risk that those who perform these functions may be either unjust, ill-informed, or unwise and that those governed may not be satisfied with the decisions made by those governing.

Behind these deliberately elementary remarks, there remains a fundamental problem, namely, the reduction of politics as such to economics.

Marx's sociology, at least in its messianic and prophetic form, presupposes the reduction of the political order to the economic order. But the political order is essentially irreducible to the economic order. Whatever the economic and social regime may be, the political problem will remain, because it consists in determining who governs, how the leaders are chosen, how power is exercised, and what the relationship of consent or dissent is between the government and the governed. The political order is as essential and autonomous as the economic order. These two orders have a reciprocal relation. The way in which production or the distribution of collective resources is organized influences the way in which the problem of authority is solved; and inversely, the way in which the problem of authority is solved influences the way in which the problems of production and of the distribution of resources are solved. The mistake is to think that a certain way of organizing production and the distribution of resources automatically solves and does away with the problem of leadership. The myth of the state's disappearance is the myth that the state exists only to produce and distribute resources and that, once this problem of production and distribution of resources is solved, there is no longer any need for a state, i.e., for leadership.[30]

This myth is doubly misleading. First, the solution of the planned economy entails a strengthening of the state. And second, even if planning did not entail a strengthening of the state, modern society would still have the problem of leadership, of the mode in which authority was to be exercised.

In other words, it is impossible to define the political regime simply by the class supposedly exercising power. The political regime of capitalism cannot be defined by the power of the monopolists any more than the political regime of a socialist society can be defined by the power of the proletariat. In the capitalist system, it is not the monopolists who personally exercise power; and in the socialist regime, it is not the proletariat which personally exercises power. In each of these two regimes, we must determine which men perform the political functions, how they are chosen, how they exercise authority, and what is the relationship of government to governed. It is impossible to reduce the sociology of political regimes to a mere appendage of the sociology of economics or of social class.

It remains for me to examine one last aspect of the problem of the relation of the economy to the whole of the society: the question of ideas or ideologies. Marx often spoke of ideas or ideologies, and he tried to explain ways of thinking—intellectual systems—in terms of their social context.

Generally speaking, in Marxist doctrine ideas belong to what Marx called the superstructure. The mode of interpretation of ideas by the reality may assume various modalities. It is possible to explain ways of thinking by the mode of production, the technical style of the society in question. But the explanation which has been most successful is the one which ascribes certain ideas to a certain social class.

In general, Marx understood by "ideology" the false consciousness or the false image a social class has of its own situation and of society as a whole. To a large extent, he regarded the theories of the bourgeois economists as a class ideology. Not that he imputed to bourgeois economists the intention of deceiving their students or their readers or of giving a false interpretation of reality; but he was inclined to think that a class cannot see the world except in terms of its own situation. As Sartre would say, the bourgeois sees the world defined by the rights he possesses in it. The juridical image of a world of rights and obligations is the social image which the bourgeois must have as a result of his situation as a bourgeois.

This interpretation—the false consciousness linked to class consciousness—can be applied to a number of ideas, ideological systems, and economic and social doctrines. But there are two difficulties about this interpretation of ideology.

First, if as a result of its situation a class has a false idea of the world—if, for example, the bourgeois class does not understand the mechanism of surplus value or remains the victim of commodity fetishism—then how did a certain individual member of this class (e.g., Marx) succeed in ridding himself of these illusions, of this false consciousness? It is possible to find an answer to this question, but then another question arises: If every class has a partial and partisan way of thinking, there is no longer any such thing as truth. How is one ideology superior to another, if every ideology is inseparable from the class that creates or adopts it? Whence the temptation to reply that, among ideologies, there is one that is superior to the others because there is one class capable of conceiving the world as it really is.

In fact, one of the tendencies of Marxist thought is to show why in the capitalist world it is the proletariat, and only the proletariat, that conceives the truth about the world, because it is only the proletariat that conceives the future beyond the revolution.

For example, if you read the works of Lukács, one of the last great Marxist philosophers, you will find in his book "History and Class Consciousness" (*Geschichte und Klassenbewusstsein*) an attempt to prove that class ideologies are not identical and that the ideology of the proletarian class is true because, in the situation imposed on it by capitalism, the proletariat and only the proletariat is capable of conceiving society in its development, in its evolution toward the revolution, and hence in its reality.[31]

There is, then, a primary theory of ideology which tries to avoid slipping into utter relativism by maintaining the link between ideology and class and at the same time the truth of one of the ideologies.

The difficulty with a formula of this kind is that the truth of this class ideology is open to debate. It is easy for defenders of other ideologies and other classes to say that we are all on the same level; assuming that our view of capitalism is governed by our bourgeois prejudice, then your proletarian view is governed by your proletarian prejudice. Why should the prejudice of the out's be better as such than the prejudice of the in's? Or, if you prefer, why should the prejudice of those on the wrong side of the tracks be better, as such, than the prejudice of those on the right side? This line of argument leads to a complete skepticism in which all ideologies are equal, equally partial and partisan, prejudiced, and therefore illusory.

Another direction has therefore been explored which seems to me preferable, the same direction that has been taken by the sociology of knowledge. There is good reason to establish distinctions between different types of intellectual constructs. In a certain sense, all thinking is related to social milieu, but the relation of painting, physics, mathematics, political economy, or political doctrines to social reality is not the same.

It is proper to distinguish scientific methods or theories related to, but not dependent on, the social reality from ideologies or misconceptions resulting from class situations which prevent men from seeing the truth. Or again, it is proper to establish distinctions between types of intellectual constructs and to study carefully the modalities of the relation of these different types of intellectual constructs to the social reality.

This task is the very one which the various sociologists of knowledge, Marxist and non-Marxist, are trying to accomplish in order to safeguard, on the one hand, the possible universal truth of certain sciences and, on the other, the possible universal value of works of art.

Indeed, it is important for a Marxist or a non-Marxist not to reduce the significance of a scientific or aesthetic production to its class content. Marx, who was a great admirer of Greek art, knew just as well as the sociologists of knowledge that the significance of human creations is not exhausted by their class content. Works of art have value and meaning even for other classes and other ages.

Without in the least denying that thinking is related to social reality and that certain varieties of thinking are related to social class, it is important to re-establish qualitative distinctions and to defend two statements which seem indispensable if we are to avoid nihilism: (1) There are domains in which the thinker can arrive at a truth valid for all, and not merely a truth of class. (2) There are domains in which the intellectual and esthetic products of societies have value and importance for the men of other societies.

I would like to conclude by indicating very briefly the Marxist schools which have come into being since Marx. Basically, there have been three major crises in Marxist thought during the last century.[32] The first is the one which has been called the crisis of "revisionism." Regrettably, this word has been used several times in history with meanings which are both different and related. The first crisis was that of German social democracy at the end of the last century and the beginning of this one. The two protagonists were Karl Kautsky and Eduard Bernstein. The basic argument was this: Is capitalist economy changing in such a manner that the revolution which we predict and on which we are counting will occur according to plan? Bernstein was the revisionist who declared that class antagonisms were not increasing in the capitalist economy, that economic concentration was not coming about either as quickly or as completely as had been anticipated, and that consequently it was not certain wisdom to rely on the historical dialectic to achieve both the catastrophe of revolution and the nonantagonistic society. This Kautsky–Bernstein dispute ended, within the German social democratic party and the Second International, with the victory of Kautsky and the defeat of the revisionists. The orthodox thesis was maintained.

The second crisis in Marxist thought was the crisis of Bolshevism. A party calling itself Marxist seized the power in Russia, and this party, as was natural, described its victory as the victory of a proletarian revolution. A fraction of the Marxists—the orthodox Marxists of the Second International, the majority of the German socialists, and the majority of Western socialists—did not agree. Since, let us say, 1917 to 1920 there has been within those parties calling themselves Marxist a dispute whose central point might be summarized as follows. Is Soviet power a dictatorship *of* the proletariat or a dictatorship *over* the proletariat? These expressions were used as early as 1917 to 1920 by the two great protagonists of this second crisis, Lenin and Kautsky. In the first crisis of revisionism, Kautsky was on the side of orthodoxy. In the Bolshevist crisis, he believed that he was still on the side of orthodoxy; but there was now a new orthodoxy to victimize him.

Lenin's thesis was simple and may be summarized as

follows. The Bolshevik party that called itself Marxist and proletarian represented the proletariat in power; the power of the Bolshevik party was the dictatorship of the proletariat. Since, after all, it had never been known with any certainty what the dictatorship of the proletariat would be like, the hypothesis that the power of the Bolshevik party was the dictatorship of the proletariat was, in the last analysis, rather tempting, and in any case it could be maintained. From this point on, everything was easy, for if the power of the Bolshevik party was the power of the proletariat, the Soviet regime was a socialist proletarian regime, and the construction of socialism followed from it.

On the other hand, according to Kautsky's thesis, a revolution occurring in a nonindustrialized country in which the working class was in a minority could not be a socialist proletarian revolution. The dictatorship of one party was not a dictatorship of the proletariat but a dictatorship over the proletariat.

From this cleavage there developed two schools of Marxist thought: one which regarded the regime of the Soviet Union as the fulfillment, with a few unforeseen modifications, of Marx's prophecies, and the other which believed that the essence of Marxist thought had been distorted, because true socialism implied not only collective ownership and planning but political democracy as well. According to the second school, socialist planning without democracy is not socialism.

The third crisis, finally, is the one in which we are living and which might be expressed by the question: Is there, between the Bolshevik version and what we might call the Scandinavian-British version of socialism, a third, intermediate term?

Today it is obvious that one of the possible modalities of a socialist society is central planning under the direction of a more or less absolute state, itself identified with a party calling itself socialist. This is the Soviet version of Marxist doctrine. There is a second version, the Western version, whose most perfect form is probably Swedish society, where there is a mixture of private and public institutions and a very extensive equalization of income. Partial planning and partially collective ownership of the instruments of production are combined with Western democratic institutions, i.e., multiple parties, free elections, free discussion of ideas and doctrines. A Sovietized socialism, on the one hand, and a bourgeoisified socialism, on the other: these are the two very obvious extremes in the modern world.

The Marxists are those who do not doubt that the true lineage of Marx is Soviet society. Western socialists do not doubt that the Western version is less unfaithful to the spirit of Marx than the Soviet version. But many Marxist intellectuals of our time find neither of these two versions altogether satisfactory. They would like a third version, that is, a society which, in a certain sense, would be as socialist and as planned as Soviet society and which, at the same time, would be as liberal as a society of the Western type. I shall not attempt to decide whether this third term can exist outside the minds of philosophers; after all, there are more things under heaven than can be found in our philosophy. Perhaps there will be a third term, but, for the moment, it is only an abstract possibility.

A last word: it is unwarranted to formulate the solution Marx would offer to our problems. One need not be a Marxist to believe that a thinker is inseparable from the age in which he lives. All those who try to prove that Marx would be a Soviet socialist or a Swedish one, symbolically speaking, are making a statement which is not only undemonstrable but to a large extent meaningless. There is no way of knowing what Marx would be today, because Marx's thought had not conceived the differentiation which the course of history has effected. Once you are obliged to say that certain phenomena which Marx criticized are imputable not to capitalism but to industrial society, others to the phase of development he observed, others solely to the system of ownership, you are embarking on a train of thought of which Marx was capable, of course (for Marx was a very great man), but which was foreign to Marx as he lived. In all probability, Marx, who had a rebel's temperament, would not be enthusiastic about any of the versions, any of the modalities of society which call themselves Marxist. But which would he prefer? An answer seems to me impossible and, in the last analysis, rather pointless.

BIOGRAPHICAL CHRONOLOGY

1818	May 5. Birth of Karl Marx, the second of eight children, in Trèves, then in Rhenish Prussia. His father, Heinrich Marx, was a lawyer and had been converted to Protestantism in 1816, though coming from a family of rabbis.
1830–35	Attended the *lycée* in Trèves.
1835–36	Studied law at the University of Bonn.
1836–41	Studied law, philosophy, and history in Berlin. Marx circulated among the young Hegelians of the Doktor Club.
1841	Received his doctorate of philosophy from the University of Jena.
1842	Marx settled in Bonn, becoming a contributor, and later a member of the editorial staff, for Cologne's *Rheinische Zeitung*.
1843	Disappointed in its shareholders' timid policies, he resigned. Married Jenny von Westphalen. Left for France. *Essay on the Jewish Question.* *Critique of Hegel's Philosophy of Law, Introduction.*
1844–45	Stay in Paris. Marx saw a lot of Heine, Proudhon, and Bakunin. Began his study of political economy. Marx recorded in several notebooks his philosophic reflections on Hegel's *Economy and Phenomenology.* He became friends with Engels. *The Holy Family* was the first book they collaborated on.
1845	Ordered by the Prussian government to leave Paris.
1845–48	Lived in Brussels. *German Ideology,* in collaboration with Engels. Misunderstanding with Proudhon. *Poverty of Philosophy* (1847). In November 1847, the Second Congress of the Communist League, which Marx attended with Engels, asked them to write a Communist manifesto.

	Publication of *The Communist Manifesto* in February 1848 in both London and Germany.
1848	Asked to leave Brussels. After spending some time in Paris, he went to Cologne where he became chief editor of *Neue Rheinische Zeitung*. Through the magazine he led an active campaign to radicalise the revolutionary movement in Germany.
1849	*Labor, Salary, and Capital* (printed in *Neue Rheinische Zeitung*).
	Marx was asked to leave the Rhineland. After another short stay in Paris, he left for London where he settled permanently.
1850	*Class Conflict in France.*
1851	Marx became a contributor to the *New York Tribune*.
1852	Dissolution of the Communist League.
	The 18th Brumaire of Louis Bonaparte.
1852–57	Marx had to give up his studies in economics in order to earn a small living from journalism. He had constant financial worries.
1859	*The Critique of Political Economy*, published in Berlin.
1860	*Herr Vogt.*
1861	Trip to Holland and Germany. Article in the Viennese magazine, *Die Presse*.
1862	Marx broke off relations with Lassalle. He had to cease writing for the New York *Tribune*. Great financial hardship.
1864	Participated in forming the International Association of Workers, for which he wrote the charter and inaugural address.
1867	Publication of Book I of *Das Kapital* in Hamburg.
1868	Marx became interested in the Russian commune and studied Russian.
1869	Disagreement with Bakunin arose during the International. Engels assured Marx of an annual income.
1871	*Civil War in France.*
1875	*Criticism of the Gotha Program.*
	Publication of the French translation of Book I of *Das Kapital*. Marx had helped the translator, J. Roy.
1880	Marx determined for Jules Guesde the grounds on which to base the French Labor Party's platform.
1881	Death of Jenny Marx. Corresponded with Vera Zassoulitch.
1882	Traveled in France and Switzerland. Stayed in Algeria.
1883	Marx died on March 14.

1885   Engel's publication of Book II of *Das Kapital*.
1894   Engel's publication of Book III of *Das Kapital*.

NOTES

1. Marx's eulogy of the revolutionary and constructive role of the bourgeoisie becomes even lyrical: "It has accomplished marvels of another order than the Egyptian pyramids, the Roman aqueducts, the Gothic cathedrals; the expeditions it has carried out are very different from the invasions and the crusades." (*Communist Manifesto*)

2. See especially Karl A. Wittfogel, *Oriental Despotism: A Comparative Study of Total Power*, New Haven, Yale University Press, 556 pp., 1957.

See also the following articles which have appeared in *Le Contrat social*: Karl A. Wittfogel, "Marx et le despotisme oriental," May 1957; Samuel H. Baron, "G. Plekhanov et le despotisme oriental," January 1959; Paul Barton, "Du despotisme oriental," May 1959, "Despotisme et totalitarisme," July 1959, and "Despotisme, totalitarisme et classes sociales," March 1960; Kostas Papaioannou, "Marx et le despotisme," January 1960.

For an orthodox Marxist consideration of this problem, see the special issue of the review *La Pensée* on "La Mode de production asiatique," No. 114, April 1964.

3. Joseph Stalin, *The Economic Problems of Socialism in the USSR* (ref.): "The main characteristics and tendencies of the fundamental economic law of modern capitalism might be formulated somewhat as follows: to assure a maximum of capitalistic profit by exploiting, ruining, and impoverishing the majority of the population of a given country, by systematically enslaving and plundering the peoples of other countries, especially those of backward countries, and finally, by provoking strikes and by militarizing the national economy with a view to insuring maximum profit. . . . The essential characteristics and tendencies of the fundamental economic law of socialism might be formulated somewhat as follows: to insure maximum satisfaction of the constantly increasing material and cultural needs of the whole society by always expanding and perfecting socialistic production on the foundation of a superior technology."

4. Indeed, it was his awareness of this incompleteness which, in addition to illness and financial difficulties, led Marx to delay publication of the last two volumes of *Capital*. From 1867 (the date of the publication of the first volume) until his death, Marx constantly pursued studies which left him unsatisfied and constantly reworked the sequel to what he regarded as his

masterwork. Thus, in September 1878, he wrote to Danielson that the second volume of *Capital* would be ready for the printer toward the end of 1879, but on April 10, 1879, he announced that he would not publish it until he had observed the development and the outcome of the industrial crisis in England.

5. The subject of the perennial decline in the rate of profit was introduced by David Ricardo and particularly developed by John Stuart Mill. Wanting to show that individuals always have motives for investing, Ricardo wrote: "There cannot then be accumulated in a country any amount of capital which cannot be employed productively until wages rise so high in consequence of the rise of necessaries, and so little consequently remains for the profits of stock, that the motive for accumulation ceases." (*Principles of Political Economy and Taxation*, London, 1819, p. 360.) In other words, for Ricardo, the decline in the rate of profit to zero is only one possibility. This decline results from an increase in the division of the product into nominal wages if the latter are forced up by a relative rise in the prices of goods necessary for survival. This rise in prices is in turn the result of the combined influence of increased demand due to demographic factors and of decreasing agricultural output. But Ricardo believed that the obstacle to growth presented by decreasing agricultural production could be removed by world trade, international specialization, and the free importation of grain from other countries. Mill, after the abolition of the Corn Laws, adopted Ricardo's theory in his *Principles of Political Economy with Some of their Applications to Social Philosophy* (1848), but he gave it a more evolutive and more long-range version which is related to modern stagnationist theories. The decline in the rate of profit is the dollars-and-cents translation on the business level of the society's advance toward a stationary state when there will no longer be any clear accumulation of capital. The law of decreasing output is the source of this decline in profit toward zero.

6. The most one can say, in an analysis of Keynesian inspiration, is that the rate of profit on the last unit of capital whose investment is necessary to maintain full employment (marginal efficacy of capital) must not be lower than the rate of interest on money as it is determined by the preference for liquidity of possessors of money. But actually, a formula of this type is difficult to integrate with Marxist economic theory, whose intellectual instruments are pre-marginalist. Moreover, there is a certain contradiction in Marx's economic analysis between the law of the falling trend of the rate of profit, which implicitly depends on the classical economists' law of outlets, and the theory of the crisis due to underconsumption on the

part of the working class, which implies a block in growth through lack of real demand. The distinction between short term and long term does not help to solve the problem, for these two theories do not propose to explain the long-term tendency on the one hand and the fluctuations on the other, but a general crisis in the economic system as a whole. (Cf. Joan Robinson, *An Essay on Marxian Economics*, London, Macmillan, 1942.)

7. In a letter to Joseph Weydemeyer dated March 5, 1852, Marx writes: "As far as I am concerned, I cannot take credit for discovering either the existence of classes in modern society or the struggle between them. Long before my time, bourgeois historians described the historical development of this class struggle and bourgeois economists discussed its economic anatomy. What I did that was new was to show: 1) that the existence of classes is associated only with certain phases of specific historical development of production; 2) that the class struggle leads necessarily to the dictatorship of the proletariat; and 3) that this dictatorship itself is only a transition to the abolition of all classes and to a classless society." (In Karl Marx-Friedrich Engels, *Etudes philosophiques*, Paris, Editions Sociales, 1951, p. 125.)

8. Alexandre Kojève, *Introduction à la lecture de Hegel*, Paris, Gallimard, 1947.

For the Marxist interpretation of Hegel, see also: Georg Lukacs, *Der junge Hegel*, Zürich-Vienna, 1948, 718 pp., and J. Hyppolite's analysis of this book in his *Etudes sur Marx et Hegel*, Paris, M. Rivière, 1955, pp. 82–104. Lukacs goes so far as to treat the subject of a theological period in Hegel as a reactionary legend and studies the criticism made by the young Hegel of the work of Adam Smith. According to Lukacs, Hegel saw the essential contradictions in capitalism without, naturally, arriving at the solution that Marx was destined to expound.

9. G. Gurvitch, *La Sociologie de Karl Marx*, Paris, Centre de documentation universitaire, 1958, typewritten, 93 pp.; *Les Fondateurs de la sociologie contemporaine, I. Saint-Simon sociologue*, Paris, Centre de documentation universitaire, 1955, typewritten, 62 pp.

10. On this problem of the relation between Marx and Saint-Simon, see also Aimé Patri, "*Saint-Simon et Marx*," *Le Contrat social*, January 1961, Vol. 5, no. 1.

11. *Phenomenology of Mind* dates from 1807. *Encyclopedia of the Philosophical Sciences* appeared in three editions during Hegel's lifetime (1817, 1827, and 1830).

12. *Grundlinien der Philosophie des Rechts*, published by

Hegel in 1821 in Berlin. This work is merely an expanded section of the *Encyclopedia*.

13. There are two texts that contain a critique of Hegel's *Philosophy of Right*. The first is the *Kritik des hegelschen Rechtsphilosophie—Einleitung*, a short text known for a long time, since it was published by Marx in 1844 in Paris in the magazine that he edited with A. Ruge, *Deutsch-französische Jahrbücher*, or *Annales franco-allemandes*. The other is the *Kritik des Hegelschen Staatsrechts, d. i. Hegels Rechtsphilosophie*, a much longer text containing a juxtalinear analysis of a part of Hegel's *Philosophy of Right*, which was not published until the 1930s, when it was published both by D. Rjazanov in Moscow on behalf of the Marx-Engels Institute, and by Landshut and Meyer in Leipzig.

On this point, see J. Hyppolite's study, "La conception hegeliene de l'Etat et sa critique par K. Marx," in *Etudes sur Marx et Hegel*, Paris, M. Rivière, 1955, pp. 120–41.

14. In *The German Ideology*, Marx writes: "From the moment labor begins to be divided, each man has an exclusive and definite sphere of activity which is imposed on him and which he cannot leave; he is a hunter, a fisherman, a shepherd, or a critic, and he must remain one if he does not want to lose his means of livelihood; whereas in the communist society, where each man does not have an exclusive sphere of activity, but can perfect himself in whatever field he likes, society regulates general production, thereby making it possible for me to do one thing today and another thing tomorrow, to hunt in the morning, fish in the afternoon, raise sheep in the evening, and practice criticism after dinner according to my whim, without ever becoming a hunter, a fisherman, or a critic. . . ." This does away with "that fragmentation of social activity, that consolidation of our own product into an objective force which dominates us, eluding our control, thwarting our expectations, reducing our calculations to nothing." (Ref.)

15. *Oekonomisch-philosophische Manuskripte*. These articles, written by Marx in Paris in 1844, remained unpublished until 1932, when they were published by D. Rjazanov in the edition Mega I, and also by S. Landshut and J. P. Meyer in a two-volume edition of Marx's writings entitled *Der historische Materialismus* (A. Kröner, Leipzig).

16. In Hegel, the three terms translated into French by *aliénation* are *Veräusserung, Entäusserung* and sometimes *Entfremdung*. For Hegel, alienation is the dialectical moment of difference or division between subject and substance. Alienation is an enriching process, and consciousness must experience numerous alienations before it can be enriched with the determi-

nations that in the end will constitute it as a totality. At the beginning of the chapter on absolute knowledge, Hegel writes: "The alienation of consciousness from the self posits thingness and this alienation has not only a negative but a positive significance, it is not only for us or in itself but for itself. . . . Such is the movement of consciousness and in the movement it is the totality of its moments. Consciousness must relate to the object according to the totality of its determination and must have apprehended it according to each of these determinations." (*Phenomenology of Mind*)

Marx gives another interpretation of alienation, for "in a certain sense totality is already given from the beginning" (J. Y. Calvez, *La Pensée de Karl Marx,* Paris, Editions du Seuil, 1956, p. 53). According to Marx, Hegel confused objectification, that is, the externalization of man in nature and the social world, with alienation. As J. Hyppolite writes, commenting on Marx: "Alienation is not objectification. Objectification is natural. It is not a way for consciousness to become alien to itself, but a way to express itself naturally." (*Logique et existence,* Paris, Presses Universitaires Françaises, 1953, p. 236.) Marx writes, "The objective being acts in an objective manner and he would not do so if objectivity were not included in the determination of his essence. He creates, he posits nothing but objects because he himself is posited by objects, because originally he is nature." (*Economic and Philosophic Manuscripts*)

This distinction, which is based on a "rational naturalism" whereby "man is directly a being of nature" (*Ibid*), allows Marx to retain, of the notion of alienation and the successive determinations of consciousness as they are expounded in *Phenomenology of Mind,* only the critical aspect. "The *Phenomenology* is a hidden criticism, still obscure to itself and mystifying, but to the degree that it retains the alienation of man—although in it man appears only in the form of mind—one finds all the elements of criticism concealed in it, and these elements are already often prepared and developed in a way that goes far beyond the Hegelian point of view." (*Ibid.,* p. 131.)

For a commentator of Hegel like J. Hyppolite, this radical difference between Hegel's and Marx's conceptions of alienation has its origin in the fact that whereas Marx begins with man as a being of nature, that is, with a positivity which is not in itself a negation, Hegel "has discovered that dimension of pure subjectivity which is nothingness" (*op. cit.,* p. 239). In Hegel, "In the dialectical beginning of history there is the limitless desire for recognition, the desire for the desire of the other, a force that is inexhaustible because it is without basic positivity" (*Ibid.,* p. 241).

17. This ambiguity in Marx's thought has been particularly brought out by Kostas Papaioannou, "La fondation du marxisme," in *Le Contrat social,* No. 6, November-December 1961, vol. 5; "L'homme total de Karl Marx," in *Preuves,* No. 149, July 1963; and *"Marx et la critique de l'aliénation,"* in *Preuves,* November 1964.

Kostas Papaioannou finds a radical opposition between the philosophy of the young Marx as it is expressed, for example, in the *Economic and Philosophical Manuscripts* and the philosophy of his maturity, especially as it is expressed in the third volume of *Capital.* For a productivist pietism which would regard labor as the exclusive essence of man and non-alienated participation in productive activity as the true end of existence, Marx substituted, according to this writer, a very classical prudence in which that human development "which alone has the value of an end in itself and which is the true reign of freedom" begins "beyond the domain of necessity."

18. This objectivist view may, moreover, be regarded as favorable or unfavorable to peace. Some say that as long as Soviet leaders are convinced of the necessary death of capitalism, the world will live in an atmosphere of crisis. But it is also possible to say, like a certain English sociologist, that as long as the Soviets continue to believe in their own philosophy, they will understand neither their society nor ours. Assured of their necessary victory, they will let us live in peace. God grant that they continue to believe in their philosophy!

19. Cf. Jean-Paul Sartre, *"Les communistes et la paix"* (*Temps modernes,* Nos. 81, 84–85, and 101), reprinted in *Situations VI,* Paris, Gallimard, 1965, 384 pp. (see also *Situations VII,* Paris, Gallimard, 1965, 342 pp. and *Critique de la raison dialectique,* Paris, Gallimard, 1960); Maurice Merleau-Ponty, *Sens et nonsens,* Paris, Nagel, 1948; *Humanisme et terreur,* Paris, Gallimard, 1947; and *Les Aventures de la dialectique,* Paris, Gallimard, 1953.

20. On the Kantian interpretation of Marxism, see: Max Adler, *Marxistische Probleme-Marxismus and Ethik,* 1913; Karl Vorländer, *Kant und Marx,* second edition, 1926; Franz Mehring, *Karl Marx, Geschichte seines Lebens,* Leipzig, 1918 (English translation: *Karl Marx, The Story of his Life,* New York, Covici Friede, 1936).

21. Lucien Goldmann, *Recherches dialectiques,* Paris, Gallimard, 1959.

22. Friedrich Engels, *Anti-Dühring.* The original German title is *Herrn Eugen Dührings Umwälzung der Wissenschaft.* The work was first published in the *Vorwärts* and the *Volksstaat* in 1877–78.

It should be noted that the *Anti-Dühring* was published during Marx's lifetime and that Marx helped his friend in the writing of it by sending him notes on various points on the history of economic thought, some of which Engels incorporated into the definitive text.

23. Lenin, *Materialism and Empiriocriticism*. In this work Lenin expounds a radical materialism and realism: "The material world perceived by the senses to which we ourselves belong is the sole reality . . . our consciousness and thought, however supra-sensible they may seem, are merely the products of a material and corporeal organ, the brain. Matter is not a product of the mind, but the mind itself is merely the superior product of matter"; or again, "The general laws of the movement of the world as well as of human thought are identical at bottom but different in their expression, in the sense that the human brain can apply them consciously, whereas in nature they move unconsciously, in the form of an external necessity, through an infinite succession of seemingly fortuitous things." This book was to become the basis of orthodox Soviet Marxism. In a letter to Gorki dated March 24, 1908, Lenin had requested the right, as a "party man," to take a position against the "dangerous doctrines," at the same time proposing to his correspondent a "pact of neutrality on the subject of empiriocriticism" which, he said, did not justify "a divisive struggle."

24. In his foreword to the *Critique of Political Economy*, Marx writes, "Neither legal relations nor the form the state takes can be explained either by themselves or by the presumed general evolution of the human mind; both have their roots in the material conditions of life," and further on, "The totality of the relations of production forms the economic structure of the society, the real foundation on which a legal and political edifice rises," or again, "The legal, political, religious, artistic, philosophical, in short, the ideological forms in which men become conscious of the conflict and carry it to the very end." (*Ibid.*)

One of the chapters in *The German Ideology* is entitled, "The Relations of the State and of Law with Property." Generally speaking, in Marx, state and law arise from the material living conditions of peoples and are the expression of the dominant will of the class that holds the power in the state.

25. Here is Marx's most significant passage, from Book III, Volume Two of *Capital*: "Formation of Shareholding Associations. Consequences: 1) The enormous extension of the scale of production and enterprises which would have been impossible to isolated capital. At the same time enterprises which were formerly governmental become associations. 2) Capital which

is based by definition on the social mode of production and presupposes a social concentration of means of production and labor force here directly assumes the form of social capital— capital of individuals directly associated—as opposed to private capital; these enterprises thus take the form of social enterprises as opposed to private enterprises. This means the abolition of capital as private property within the limits of the capitalist mode of production itself. 3) The transformation of the truly active capitalist into a mere director and administrator of the capital of somebody else and of proprietors of capital into mere proprietors, mere capitalist financiers. . . . This amounts to the abolition of the capitalist mode of production within the capitalist mode of production itself, hence a contradiction which destroys itself and which by all evidence appears as a mere transitional phase pointing to a new form of production. This transitional phase appears as a similar contradiction. In certain spheres it establishes a monopoly, thus provoking the interference of the state. It gives rise to a new financial aristocracy, a new species of parasites in the form of purely nominal planners, founders, and directors; a whole system of swindling and fraud with respect to the floating, issuing, and trading of shares. This amounts to private production without the control of private ownership." In Marx, the critic and even the pamphleteer is never far from the economic analyst and the sociologist.

26. *Capital,* Book III, Chapter 52. Marx continues, "It is unquestionably in England that the economic division of modern society finds its most advanced and most classic development. Nevertheless, even in this country the division into classes does not appear in pure form. Here, too, the intermediary and transitional stages blur the precise demarcations (albeit much less in the country than in the towns). For the purposes of our study, however, this is not important. We have seen that the capitalist mode of production has a perennial tendency—this is the law of its evolution—increasingly to separate means of production and labor and to concentrate these scattered means of production into large groups, thus transforming labor into salaried labor and means of production into capital. From another point of view, this tendency has for a corollary the separation of landed property which becomes autonomous in relation to capital and labor, or else the transformation of all landed property into a form of property corresponding to the capitalist mode of production.

"The question that arises from the outset is, What constitutes a class? The answer flows quite naturally from the answer to another question, Why do the salaried workers, the capitalists, and the landowners constitute the three great classes of society?

"On first sight the answer would seem to be an identity of income and of sources of incomes. Here we have three important social groups whose members live respectively on wages, profit, and ground rent; on the renting of their labor force, their capital, and their landed property.

"However, according to this definition, doctors and civil servants, for example, would also constitute two distinct classes, for they belong to two distinct social groups whose members derive their income from a single source. This distinction would also apply to the infinite variety of interests and situations caused by the division of social labor within the working class, the capitalist class, and the landowning class, the latter, for example, being divided into vinegrowers, owners of fields, of forests, of mines, of fisheries, etc." (*Here the manuscript breaks off* [Friedrich Engels].)

27. *The Class Struggle in France* (*1848–1850*). Written between January and October 1850, this text, which did not appear in a pamphlet and under this title until 1895, is composed mainly of a series of articles which appeared in the first four issues of the *Neue Rheinische Zeitung,* an economic and political review whose publication began in London in March 1850.

The Eighteenth Brumaire of Louis Bonaparte. Written between December 1851 and March 1852, this text was published for the first time in New York on May 20, 1852 by Weydemeyer.

28. Marx writes, in Volume I of *Capital:* "The fanatical agent of accumulation, the capitalist, forces men, without mercy or respite, to produce for the sake of producing, and thus instinctively drives them to develop the productive powers and the material conditions which alone can form the foundation for a new and better society. The capitalist is respectable only insofar as he is capital in human form. In this role he is also like the miser, dominated by his blind passion for abstract wealth, value. But what in the miser seems to be a personal mania is in the capitalist the effect of a social mechanism in which he is only a cog. The development of capitalist production necessitates a continual increase in the capital invested in an enterprise, and competition imposes the immanent laws of capitalist production as coercive laws external to each individual capitalist. Competition does not permit him to save his capital without increasing it, and he cannot continue to increase it except by progressive accumulation." Or again, "Save, always save, that is, constantly reconvert into capital the largest possible share of surplus value or net product! Accumulate for the sake of accumulating, produce for the sake of producing, this

is the watchword of the political economy that proclaims the historic mission of the bourgeois period. And it has no illusions about the labor pains of wealth: but what is the point of jeremiads which in no way alter historic fatalities? From this point of view, if the proletariat is nothing but a machine for producing surplus value, the capitalist is nothing but a machine for turning this surplus value into capital."

29. Marx writes, "Between capitalist society and communist society there is a period of revolutionary transformation. This period also corresponds to a transitional political phase in which the state can be only the revolutionary dictatorship of the proletariat." (*Capital,* Vol. I, p. 1429.) Marx also uses this expression in the letter to Joseph Weydemeyer (March 5, 1852) quoted in Note #7, and the idea, if not the phrase, is already found in the *Communist Manifesto:* "The proletariat will make use of political power to appropriate gradually every kind of capital from the bourgeoisie, to centralize all instruments of production in the hands of the state—that is, of the proletariat organized into the ruling class—and to increase the mass of productive forces as rapidly as possible."

On the frequency of the use of the term "dictatorship of the proletariat" by Marx and Engels, see Karl Draper, "Marx and the Dictatorship of the Proletariat," *Cahiers de l'I. S. E. A.,* Series S, No. 6, November 1962.

30. Marx shared this devaluation of the order of politics, which is reduced to economics, with Saint-Simon and the Manchesterian liberals. In *L'Organisateur* (Vol. 4, pp. 197–98), Saint-Simon had written, "In a society organized for the positive aim of working for its prosperity by the sciences, the fine arts, and the arts and crafts," as opposed, therefore, to military and theological societies, "the most important act, the one that determines the direction in which the society is to advance, is no longer that of men invested with governmental functions, but is performed by the social body itself. Moreover, the end and object of such an organization are so clear and so well defined that there is no longer any room for arbitrariness in men or even in laws. Under an order of this kind, the citizens responsible for the different social functions, even the most elevated ones, in a sense perform only subordinate roles, since their functions, no matter how important, consist only in marching in a direction which has not been chosen by them. The action of governing is then non-existent, or almost non-existent as far as signifying the action of commanding." (Passage quoted by G. Gurvitch in his *Cours sur les fondateurs de la sociologie contemporaine: Saint-Simon,* p. 29.)

On the political thought of Marx, see Maximilien Rubel, "Le

concept de démocratie chez Marx," in *Le Contrat social*, July-August 1962; Kostas Papaioannou, "Marx et l'Etat moderne," in *Le Contrat social*, July 1960.

31. Georg Lukacs, *Geschichte und Klassenbewusstsein*, Berlin, 1923.

32. For a more detailed analysis, cf. my study, "L'impact du marxisme au xxe siècle," *Bulletin S. E. D. E. I. S., Etudes,* No. 906, January 1, 1965.

Alexis de Tocqueville

I

TOCQUEVILLE is not ordinarily included among the founders of sociology; I consider this neglect of Tocqueville's sociological writings unjustified. But I have still another reason for wishing to discuss him. For in studying Montesquieu, Comte, and Marx, the relation between economic phenomena and the political organization of society was necessarily central to my analyses. Each time I explained first the interpretation given by the thinker of the society in which he was living. The historical diagnosis was the primary fact to which I related the system of each sociologist. Tocqueville's historical diagnosis differs from that of Comte or Marx. Instead of giving priority either to the industrial reality, as Comte did, or to the capitalist reality, as Marx did, he gave priority to the democratic reality. I shall attempt to define what Tocqueville meant by the irresistible advance of modern societies toward democracy. And I shall explore the manner in which Tocqueville himself conceived his work and, as we might say in a language unknown to him, the manner in which he conceived sociology.

Tocqueville's theory, if we may call it that, involves the determination of certain structural traits of modern societies and then the comparison of the various modalities of these societies. Now, Comte observed industrial society, and without in the least denying that it may admit of secondary differences from one nation or continent to another, he put the emphasis on the characteristics common to all industrial societies. Comte defined industrial society and believed that from this definition one could indicate the characteristics of the political and intellectual organization of every industrial society. Marx defined the capitalist regime and believed that from this definition one could determine certain phenomena which must exist in all societies with capitalist regimes. Tocqueville, on the other hand, noted certain characteristics which are related to the essence of any modern or democratic society; but he added that, beyond these foundations common to all modern societies, there are many possible political regimes. Democratic societies may be liberal or despotic; democratic societies may and must assume different forms in the United States or in Europe, in Germany or in France. Consequently, Tocqueville is a comparative sociologist par excellence; he tries to determine significance by comparing types of societies belonging to the same species.

Now, since I personally consider the essential task of sociology to be precisely this comparison of types within the same species, I feel it is worthwhile to set forth briefly the leading ideas of a man who in Anglo-Saxon countries is regarded as one of the greatest political thinkers of the nineteenth century, the equal of Montesquieu in the eighteenth—and yet who, in France, has always been neglected by sociologists. The modern sociological school of Durkheim derived from the work of Auguste Comte, and because of this, French sociologists have emphasized the phenomena of social structure at the expense of the phenomena of political institutions. This is probably the reason Tocqueville has not figured in the foreground of their thinking.

Tocqueville wrote two major books, of which one is *Democracy in America*. He wrote it when very young, after a trip to America, and it immediately had an extraordinary commercial success and an intellectual success as well. He published the first volume in 1835 and the second five years later. He was a deputy under the regime of Louis Philippe. He himself belonged to a family of the old Norman nobility and was elected in the district where his château was located and still stands today. At the assembly of the Orleanist monarchy, he was ranked among what was called the "dynastic opposition." He had little liking for the regime of Louis Philippe, which he considered narrow, egoistical, a monopoly of the middle class which had secured power and places for itself. In a famous speech of January 1848, he predicted the coming of the revolution, although he considered this revolution one of the most senseless in the history of France. Under the Second Republic, he was for five months minister of foreign affairs to Louis Bonaparte, President of the Republic. The head of his cabinet was Count Gobineau, the racist theoretician, for whom Tocqueville had some affection but whose ideas he did not share. (A very interesting correspondence between Gobineau and Tocqueville has been published.) After the *coup d'état* of Louis Bonaparte, Tocqueville retired from politics and began to study the origins of nineteenth-century France. Somewhat as Taine, after the defeat of 1870, studied the origins of contemporary France, Alexis de Tocqueville, after the *coup d'état,* studied the origins of the Revolution, and his efforts resulted in a book entitled *The Old Regime and the French Revolution.* This book constitutes only the first part of a general work which he intended to devote to the French Revolution and the formation of modern French society. His model was Montesquieu's *Considerations on the Causes of the Grandeur and Decadence of the Romans.* It was also his ambition to make history intelligible—not to recount events, but to determine their underlying causes. He did not have time to write the second part of the book he contemplated. Fragments of it, found among his papers, have been published.

Besides these two principal books, *Democracy in America* and *The Old Regime and the French Revolution,* there appeared after his death a volume called *Recollections—* recollections of the Revolution of 1848 and of his time spent at the ministry of foreign affairs. His correspondence and speeches have also been published. But most important to us are his two major books, one on America and the other on France, which are, as it were, the two leaves of a diptych. The book on America seems to have been intended to answer the question: Why is American democratic society liberal? *The Old Regime and the French Revolution* is intended to answer the question: Why does France, in the course of her evolution toward democracy, have so much trouble maintaining a political regime of liberty?

The point of departure for our study must necessarily be the definition of the notion of democracy or democratic society which runs through Tocqueville's writings, just as with Comte we defined the notion of industrial society, or with Marx the notion of the capitalist regime. This definition of a democratic society is not reached without some difficulty, and it has often been said that Tocqueville used the word constantly without ever defining it accurately.

I should like to quote first a passage from the second volume of *The Old Regime and the French Revolution*, i.e., from the collection of notes found among Tocqueville's papers.

> Much confusion is caused by the employment given to these words: *democracy, democratic institutions, democratic government*. Unless they are clearly defined and unless there is agreement about their definition, we shall live in an inextricable confusion of ideas, to the great advantage of demagogues and of despots:
>
> They will say that a country governed by an absolute ruler is a *democracy* because he governs by such laws and maintains such institutions as are favorable to the great mass of the people. Such a government, it will be said, is *democratic*, a *democratic monarchy*.
>
> But *democratic government, democratic monarchy* can mean only one thing in the true sense of these words: a government where the people more or less participate in their government. Its sense is intimately bound to the idea of political liberty. To give the democratic epithet to a government where there is no

political liberty is a palpable absurdity, since this departs from the natural meaning of these words.

Such false or obscure expressions are adopted: *a*) because of the wish to give the masses illusions, since the expression "democratic government" will always evoke a certain degree of appeal; *b*) because of the embarrassing difficulty in finding a single term which would explain the complex system of an absolute government where the people do not at all participate in public affairs but where the upper classes have no privileges either and where legislation aims to provide as much material welfare as possible.

I am inclined to say that this passage, which I selected deliberately, gives democracy a peculiarly political definition which is not particularly consistent with the way in which Tocqueville ordinarily used the term. Indeed, in this passage Tocqueville seems to imply that in reality democracy exists only to the extent that political liberty does. In this case, the word *democracy* would still have its traditional acceptation. It would designate a mode of government; and if this mode of government were not characterized by popular participation in the exercise of power, there would be no democracy.

But, as a matter of fact, more often than not Tocqueville did not use the word *democracy* in this sense, but to designate a certain type of society much more often than a certain type of power. To give a counterpart to the first passage I quoted, I shall quote another, this time from *Democracy in America*.

But if you hold it expedient to divert the moral and intellectual activity of man to the production of comfort and the promotion of general well-being; if a clear understanding be more profitable to man than genius; if your object is not to stimulate the virtues of heroism, but the habits of peace; if you had rather witness vices than crimes, and are content to meet with fewer noble deeds, provided offenses be diminished in the same proportion; if, instead of living in the midst of a brilliant society, you are contented to have prosperity around you; if, in short, you are of the opinion that the principal object of a government is not to confer the greatest possible power and glory upon the body of the nation, but to ensure the greatest enjoyment and to avoid the most misery to each of the individuals who compose it—if such be your desire, then equalize the conditions of men and establish democratic institutions.

But if the time is past at which such a choice was possible, and if some power superior to that of man already hurries us, without consulting our wishes, towards one or the other of these two governments, let us endeavor to make the best of that which is allotted to us and, by finding out both its good and its evil tendencies, be able to foster the former and repress the latter to the utmost.

This highly eloquent passage, full of rhetorical antitheses, is altogether characteristic of the style, the manner, and, I would even say, the substance of Tocqueville's thought.

For Tocqueville, democracy is the equalization of conditions. That society is democratic in which there no longer exist, as in the old regime, distinctions of orders and classes, in which all the individuals who make up the collectivity are socially equal—I repeat, socially equal: this does not mean intellectually equal, which would be absurd, or economically equal, which, according to Tocqueville, would be impossible. To be socially equal means that there is no hereditary difference of conditions and that all occupations, all professions, all titles, and all honors are accessible to all.

But if such is the essence of democracy, it is clear that the government proper to an equalitarian society is the one which, in the first quotation, Tocqueville called democratic government. For if there is no essential difference of condition between members of the collectivity, it is natural that the sovereignty be in the hands of the people as a whole. Thus we return to the definition of democracy given by Montesquieu and the classical authors. The whole of the social body is sovereign, and participation of all in the choice of rulers and in the exercise of authority is the logical expression of a democratic society, that is to say an equalitarian society.

But further, as one learns from the second passage, a society of this order, in which equality is the social law and democracy the temper of the state, is also a society whose first objective is the well-being of the greatest number. It is a society whose objective is not power or glory, but peace and prosperity. If we were to use the language of today, we would call it a bourgeois society. And Tocqueville, as the descendant of a noble family, fluctuated in his judgment of democratic society between severity and indulgence, between the reticence of his heart and the reluctant consent of his reason. (I am paraphrasing a letter of Tocqueville's in which he wrote that his reason favors a society of this kind, whose objective and justification is to assure the greatest well-being of the greatest number, but that his heart cannot defend without reservations a society in which the sense of grandeur and glory tends to disappear.)[1]

If this is the characteristic of modern democratic society, I think the central problem in Tocqueville may be understood by referring to the author who he himself said was constantly in his mind when he was writing *Democracy in America,* namely, Montesquieu. For this central problem in Tocqueville is a development of one of the problems raised by Montesquieu.

One may recall that according to Montesquieu, two of his three political regimes (republic, monarchy, and despotism) were moderate and the third was not. Republic and monarchy are, or can be, moderate regimes in which liberty is preserved; while, by definition, despotism, or the arbitrary rule of one man, is not and cannot be a moderate regime. Between the two moderate regimes, republic and monarchy, we observe a fundamental opposition: between the equality of orders and conditions which is the principle of the ancient republics and the inequality of orders and

243

conditions which is the essence of modern monarchies, or at any rate of the French monarchy. As a consequence, Montesquieu thought that liberty can be preserved in political regimes in two ways or in two types of society: the small republics of antiquity, whose principle is virtue and in which individuals are and must be as equal as possible; and, on the other hand, the modern monarchies, large states whose principle is honor and in which inequality of conditions is, as it were, the very condition of liberty. In fact, it is to the degree that each man feels obliged to remain loyal to the obligations of his rank that the power of the king, which must be exercised according to law and with moderation, does not degenerate into absolute and arbitrary power.

In other words, in the modern monarchy—at least in the French monarchy as conceived by Montesquieu—inequality is both the mechanism and the guarantee of liberty. But in studying England, Montesquieu had studied the phenomenon, new to him, of representative government. He had discovered in England that the aristocracy engaged in commerce and was not corrupted thereby. There he had observed a liberal, commercial monarchy in operation.

Tocqueville's conception might be regarded as the development of Montesquieu's theory of the English monarchy. Tocqueville, writing after the Revolution, could not imagine that for moderns the foundation and guarantee of liberty is inequality of conditions, for the simple reason that inequality of conditions had disappeared, at least in his eyes. It would be senseless to wish to restore the authority and privileges of an aristocracy which was destroyed by the Revolution.

The liberty of the moderns, to speak in the manner of Benjamin Constant, cannot therefore be based, as Montesquieu suggested in his theory of monarchy, on the differentiation of orders and states. Equality of condition is the major fact, an irresistible development which must inspire us with a sort of sacred terror, as before a movement willed by Providence. Again I am paraphrasing Tocqueville; read the Preface to *Democracy in America*.[2] Here Tocqueville's point is this: since the liberty of the moderns cannot be based on inequality of condition, it must be based on the democratic reality of equality of condition. But this liberty can be safeguarded in an equalitarian society only by institutions whose model he believed he had found in America.

What does he understand by liberty? Again, it must be understood at once that Tocqueville, who did not write in the manner of modern sociologists, had not given a careful definition of what he meant by liberty. But, in my opinion, it is not difficult to state this precisely, according to the requirements of twentieth-century pedantry. Moreover, I think his conception of liberty closely resembles Montesquieu's.

The first term that constitutes the content of "liberty" is "security," that is, the guarantee against arbitrary governments. When power is exercised only according to law, individuals have security. But since men are not to be trusted and since no individual is virtuous enough to wield absolute power without becoming corrupted—again I am paraphrasing Tocqueville—absolute power cannot be given to anyone. It is necessary, therefore, as Montesquieu would have said, for power to check power. There must be a plurality of centers of force—a plurality of political and administrative organs which balance one another. And since all men participate in the sovereignty, those who exercise power must, in a sense, be the representatives or delegates of the governed. In other words, there must be self-government; the people, insofar as materially possible, must govern themselves.

Hence, the problem in Tocqueville might be summarized as follows: Under what conditions can a society in which the lot of individuals tends to become uniform avoid falling into despotism and safeguard its liberty? In ordinary language, one might say that the problem in Tocqueville is that of the compatibility of equality and liberty.

Before commenting further on *Democracy in America*,

perhaps I should say a word about Tocqueville's interpretation of what was essential in the eyes of his contemporaries, Comte and Marx.

To my knowledge, Tocqueville did not know the work of Auguste Comte; he must have heard of it, but it does not seem to have played any sort of role in the development of his thought. As for Marx's work, my guess is that he did not know it. Tocqueville could, of course, have read *The Communist Manifesto;* but *The Communist Manifesto* was more famous in 1948 than it was in 1848. In 1848 it was the pamphlet of a political refugee who had fled to London; and there is no evidence, to my knowledge—though I say it with hesitation, for one would have to read or reread all of Tocqueville's correspondence in order to be sure—that Tocqueville knew of this obscure pamphlet which has since become so famous.

As for those phenomena which were essential in the eyes of Comte and Marx, i.e., industrial society and capitalism, naturally Tocqueville did discuss them. He agreed with Comte as well as with Marx on the apparently obvious fact that the preferred activity of the societies of his day was commercial and industrial activity; he said so apropos of America, and he assumed that the same tendency prevailed in European societies. Although he did not express himself in the same style as Saint-Simon or Auguste Comte, he would also have been inclined to contrast the societies of the past, in which military activity predominated, with the societies of his time, whose objective and mission was the well-being of the greatest number.[3]

But Tocqueville had a tendency to combine commercial and industrial activity; and when he discussed this predominance of commercial and industrial activity in American society, or even in European societies, he interpreted it primarily in relation to the past and to his central theme, that of democracy. He tried to show that industrial and commercial activity do not re-establish an aristocracy in the manner of the past. The inequality of fortunes implied by commercial and industrial activity did not seem to him to contradict the equalitarian tendency of modern societies.

First of all, in his eyes, commercial, industrial, transferable wealth is mobile, so to speak. It does not become crystallized in families that maintain their privileged position from generation to generation.

Further, he believed that between the captain of industry and his workers there do not arise those hierarchical ties of dependence which formerly existed between the aristocrat and the peasants or farmers. He regarded the only historical foundation for a true aristocracy to be land-ownership or military activity.

Consequently, in his sociological vision, inequalities of wealth, however extreme, do not contradict the fundamental equality of condition characterizing modern societies. There is a passage in Chapter XX of *Democracy in America* in which, speaking of the captains of industry, for a moment Tocqueville suggested that if an aristocracy is ever to be re-established in democratic society, it will be by the captains of industry.[4] But, generally speaking, it may be said that he did not believe that modern industry would give rise to an aristocracy. He believed that inequalities of wealth would tend rather to diminish in proportion as modern societies become more democratic, and above all he believed that these industrial and commercial fortunes were too precarious to be the basis for a durable hierarchical structure.

In other words, it may be said that in distinction to that catastrophic and apocalyptic vision of the development of capitalism characteristic of Marx's thought, Tocqueville, as early as 1835, was fashioning the half-enthusiastic, half-resigned—perhaps more resigned than enthusiastic—theory of what is known today as the welfare state, or universal *embourgeoisement*.

It is interesting to compare the three visions of Comte, Marx, and Tocqueville. One was the organizational vision of those we today call technocrats; the next, the apocalyptic vision of those who yesterday were revolutionaries; the third was the equable vision of a society in which everyone has something and everyone, or almost everyone, is interested in maintaining the social order. Personally, I think that of these three visions the one that most closely resembles Western European societies today is Tocqueville's. In all fairness, I should add that the European society of the 1930's tended to be closer to Marx's vision; but which of these visions will most closely resemble European society in the 1970's and 1980's is a question which, of course, no one is now in a position to answer.

Let us turn now to *Democracy in America,* bearing in mind the question raised by Tocqueville: Why is American democracy free or liberal?

I shall quote Tocqueville's own enumeration of causes because it also indicates his theory of determinants or variables. The three kinds of causes named by Tocqueville are (1) the accidental and particular situation in which American society happened to be, (2) the laws, and (3) the customs and manners.

The accidental and particular situation means the geographical space where the European immigrants settled and the absence of neighboring states, i.e., enemy states, or at any rate potential enemies. Up to the time Tocqueville was writing, American society had enjoyed the exceptional advantage of having a minimum of diplomatic obligations and of running a minimum of military risks. At the same time, it was a society created by men provided with all the technical equipment of a mature civilization, who had settled on an unlimited space—a situation which was unparalleled in Europe and which, in Tocqueville's eyes, afforded one explanation for both the absence of an aristocracy and the preeminence accorded to industrial activity.

It might be said, adopting a theory of modern sociology, that the condition for the formation of a landed aristocracy is scarcity of land. But if scarcity of land is the condition for aristocratic ownership, it is easily understood why the latter did not come into being in the American vastness. The idea is found in Tocqueville, but it is only one among many, and I do not assume that for him it was the fundamental explanation.

He was more inclined to put the emphasis on the system of values of the Puritan immigrants, with their double sense of equality and liberty; and he outlined a theory according to which the characteristics of a society proceed, as it were, from its origins. Thus, American society has supposedly preserved the characteristics of its founders, the first immigrants.

Let me add that Tocqueville, as a good student of Montesquieu, established a hierarchy among the three kinds of causes he named: geographical and historical situation exert less influence than laws, and laws are less important than customs, manners, and religion. In the same conditions, with other manners and other laws, another society would have emerged. The historical and geographical conditions he analyzed were merely favorable circumstances. The true causes of the liberty enjoyed by American democracy are good laws and even more so the customs, manners, and beliefs without which there can be no liberty.

Thus, American society can provide, not a model, but a lesson to European societies by showing them how liberty is safeguarded in a democratic society.

II

AT THE END of the last section, I indicated briefly what Tocqueville meant by the first category of causes, that is to say, the origins and mentality of the founders, the immensity of the territory, the abundance of resources. This brings me to the second category of causes, namely, the laws.

The chapters Tocqueville devoted to American laws may be studied from two points of view. On the one hand, we may ask to what extent Alexis de Tocqueville accurately understood the functioning of the American Constitution in his time and to what extent he foresaw the changes that have taken place in it. In other words, there is a possible, interesting, and legitimate study which would consist of a comparison of Tocqueville's interpretation with interpretations which were given in his time or are still given today. There is, indeed, a considerable American literature on this subject. In particular, one American historian, G. W. Pierson, has reconstructed Tocqueville's journey, detailed the traveler's encounters with important Americans, and discovered the origins of certain of his ideas; in other words, he has compared Tocqueville, as interpreter of American society, with his informants and commentators.[5] I shall not be concerned here with this sort of study.

The second possible method consists simply in tracing the broad outlines of Tocqueville's interpretation of the American Constitution and showing its relation to general sociological problems, Montesquieu's and his own: Which are the laws, in a democratic society, most propitious to the safeguarding of freedom?

Here, very generally, are Tocqueville's essential ideas. Tocqueville emphasized the benefits which the United States derives from the federal character of its Constitution. A federal constitution may, in a certain sense, combine the advantages of both great and small states. (It may be recalled that Montesquieu, in *The Spirit of the Laws,* also devoted a few chapters to this same principle, which makes it possible to organize the power necessary to the security of the state without incurring the evils peculiar to great concentrations of men.) I shall quote a passage from the first volume of *Democracy in America.*

If none but small nations existed, I do not doubt that mankind would be more happy and more free; but the existence of great nations is unavoidable.

Political strength thus becomes a condition of national prosperity. It profits a state but little to be affluent and free if it is perpetually exposed to be pillaged or subjugated; its manufactures and commerce are of small advantage if another nation has the empire of the seas and gives the law in all the markets of the globe. Small nations are often miserable, not because they are small, but because they are weak; and great empires prosper less because they are great than because they are strong. Physical strength is therefore one of the first conditions of the happiness and even of the existence of nations. Hence it occurs that, unless very peculiar circumstances intervene, small nations are always united to large empires in the end, either by force or by their own consent. I do not know a more deplorable condition than that of a people unable to defend itself or to provide for its own wants.

The federal system was created with the intention of combining the different advantages which result from the magnitude and the littleness of nations; and a glance at the United States of America discovers the advantages which they have derived from its adoption.

In great centralized nations the legislator is obliged to give a character of uniformity to the laws, which does not always suit the diversity of customs and of districts; as he takes no cognizance of special cases, he can only proceed upon general principles; and the population are obliged to conform to the requirements of the laws, since legislation cannot adapt itself to the exigencies and the customs of the population, which is a great cause of trouble and misery. This disadvantage does not exist in confederations.

We see that, in this passage, Tocqueville revealed a certain pessimism as to the possibility of the existence of small nations which are not strong enough to defend themselves. It is curious to reread this passage today, for we wonder what, given this view of human affairs, Tocqueville would say about the large number of nations, incapable of defending themselves, which are being born in the world right now. Perhaps, however, he would revise this general formula by adding that nations which are not self-sufficient may be capable of surviving if the conditions necessary to their security are created by the international system.

However this may be, Tocqueville—true, as it were, to the constant conviction of the classical philosophers—required the state to be large enough to possess the power necessary to security and small enough for legislation to be adapted to the diversity of circumstances and localities. This combination obtains only in a federal or confederate constitution, which, in Tocqueville's eyes, was the first and highest merit of the laws that Americans have made for themselves.

Further, he saw quite clearly that the American federal Constitution guaranteed what is known today as the free circulation of property, persons, and capital; in other words, that the federal principle would be able to prevent the formation of internal customs duties and hence the dislocation of the economic unity of the American territory.

Tocqueville believed that two principal dangers threaten the existence of democracies. They are "the complete subjection of the legislature to the will of the electoral body, and the concentration of all the other powers of the government in the legislative branch." These two dangers are expressed in traditional language. A democratic government, in the eyes of a Montesquieu or a Tocqueville, must not be such that the people can abandon themselves to all the impulses of their passions and determine the decisions of the government. Moreover, Tocqueville held that every democratic regime tends toward centralization and simultaneously toward the concentration of power in the legislative body.

Now, the American Constitution provides for the division of the legislative body into two assemblies. It establishes a presidency of the republic, which Tocqueville in his day considered weak, but which is relatively independent of direct pressures from either the electoral or the legislative body. Moreover, in the United States the legal spirit is the only substitute for an aristocracy and the respect for legal form is favorable to the safeguarding of freedom. Tocqueville noted further the plurality of American political parties which, moreover, he rightly remarked, are not, like French parties, motivated by ideological convictions; they do not defend contradictory principles of government but represent the organization of interests and tend merely to discuss pragmatically the problems facing society.

Tocqueville added two other semiconstitutional political circumstances which contribute to the safeguarding of liberty. One is the freedom of association and the increase of voluntary organizations. As soon as a question is raised in a town, a county, or even on the level of the federal state as a whole, a certain number of citizens appear and form a voluntary organization for the purpose of studying and possibly answering it. To paraphrase Tocqueville, whether it is a case of building a hospital in a small town or of investigating the cause of war or abolishing war, whatever its order of magnitude, a voluntary organization will emerge to devote its free time and especially its money to studying the problem.

The second semiconstitutional political circumstance is freedom of the press. Tocqueville said, in essence, that it is burdened with disadvantages of all kinds, so greatly are the newspapers inclined to abuse it, so difficult is it for such freedom not to degenerate into license. But he continued, in a formula resembling Churchill's statement about democracy, to the effect that there is only one regime worse than freedom of the press, and that is the suppression of that freedom. In the societies of today, total freedom is still preferable to total suppression of that freedom. And Tocqueville seemed to think that between these two extreme forms there is scarcely any middle ground.

One may study with profit the many pages Tocqueville devoted to the American legal system, the jury, the legal and even political function of the jury. But here I shall proceed immediately to the third category of causes, that is, to manners and beliefs, and above all to Tocqueville's central idea—central to his interpretation of American society but also to the comparison he was constantly making, explicitly or implicitly, between America and Europe.

This fundamental theme is that in the last analysis freedom depends on the manners and beliefs of the men who are to enjoy it. The decisive factor in these manners is religion. American society was, in Tocqueville's eyes, the society able to combine the spirit of religion and the spirit of liberty; and were we to seek a single reason why in America the survival of liberty is probable while in France its future is precarious, the answer, according to Tocqueville, would be that American society combines the spirit of religion and the spirit of liberty, while French society is torn by the opposition between church and democracy, or religion and liberty.

We might say that, according to Tocqueville, it is the conflict in France between the modern spirit and the church which underlies the difficulties democracy encounters in remaining liberal and that on the other hand the kinship of inspiration between the spirit of religion and the spirit of liberty is the ultimate foundation of American society. I shall quote two passages on this essential subject.

> I have said enough to put the character of Anglo-American civilization in its true light. It is the result (and this should be constantly kept in mind) of two distinct elements, which in other places have been in frequent disagreement, but which the Americans have succeeded in incorporating to some extent one with the other and combining admirably. I allude to the *spirit of religion* and the *spirit of liberty*.
>
> The settlers of New England were at the same time ardent sectarians and daring innovators. Narrow as the limits of some of their religious opinions were, they were free from all political prejudices.
>
> Hence arose two tendencies, distinct but not opposite, which are everywhere discernible in the manners as well as the laws of the country.

And a little further on he concluded:

> Thus in the moral world everything is classified, systematized, foreseen, and decided beforehand; in the political world everything is agitated, disputed, and uncertain. In the one is a passive though a voluntary obedience; in the other, an independence scornful of experience, and jealous of all authority. These two

tendencies, apparently so discrepant, are far from conflicting; they advance together and support each other.

Religion perceives that civil liberty affords a noble exercise to the faculties of man and that the political world is a field prepared by the Creator for the efforts of mind. Free and powerful in its own sphere, satisfied with the place reserved for it, religion never more surely establishes its empire than when it reigns in the hearts of men unsupported by aught beside its native strength.

Liberty regards religion as its companion in all its battles and its triumphs, as the cradle of its infancy and the divine source of its claims. It considers religion as the safeguard of morality, and morality as the best security of law and the surest pledge of the duration of freedom.

Apart from the style, which is not one we would use today, this passage strikes me as an admirable sociological interpretation of the manner in which, in a civilization of the Anglo-American type, religious severity and political liberty may be combined. A modern sociologist would translate these phenomena into more refined, more complicated concepts; but it does not seem absolutely certain to me that such complications would add very much to the idea.

In any case, as a sociologist Tocqueville was still in the tradition of Montesquieu: he wrote in the language of all, he is comprehensible to all, and he was more concerned with giving the idea a literary form than with multiplying concepts or distinguishing criteria.

By way of complement and contrast, I should now like to quote another passage from the first volume of *Democracy in America,* in which Tocqueville explained that the relation of religion and liberty in France is the exact opposite of what it is in the United States.

If any hold that the religious spirit which I admire is the very thing most amiss in America, and that the only element wanting to the freedom and happiness of

Main Currents in Sociological Thought

the human race on the other side of the ocean is to believe with Spinoza in the eternity of the world, or with Cabanis that thought is secreted by the brain, I can only reply that those who hold this language have never been in America and that they have never seen a religious or a free nation. When they return from a visit to that country, we shall hear what they have to say.

There are persons in France who look upon republican institutions only as a means of obtaining grandeur; they measure the immense space that separates their vices and misery from power and riches, and they aim to fill up this gulf with ruins, that they may pass over it. These men are the *condottieri* of liberty, and fight for their own advantage, whatever the colors they wear. The republic will stand long enough, they think, to draw them up out of their present degradation. It is not to those that I address myself. But there are others who look forward to a republican form of government as a tranquil and lasting state, towards which modern society is daily impelled by the ideas and manners of the time, and who sincerely desire to prepare men to be free. When these men attack religious opinions, they obey the dictates of their passions and not of their interests. Despotism may govern without faith, but liberty cannot. Religion is much more necessary in the republic which they set forth in glowing colors than in the monarchy which they attack; it is more needed in democratic republics than in any others. How is it possible that society should escape destruction if the moral tie is not strengthened in proportion as the political tie is relaxed? And what can be done with a people who are their own masters if they are not submissive to the Deity?

This passage, which I find admirable, is also typical of the third party in France, the one which will never be strong enough to exercise power, because it is at the same time democratic, favorable or resigned to representative institutions, and hostile to the antireligious. Tocqueville was a liberal who would like to have seen the democrats recognize the necessary dependence between free institutions and religious beliefs. But, as a result of his knowledge of history and his sociological analyses, he must have known (and probably did know) that this reconciliation was improbable. The conflict between the Catholic Church and the modern spirit in France has a long tradition, as does the amity between religion and democracy in Anglo-American civilization. Hence Tocqueville simultaneously had to deplore the conflict and reveal its causes, which are difficult to eradicate, as is proved by the fact that over a century after Tocqueville's time the conflict is not yet resolved.

Let us return to our fundamental theme, the necessity, in an equalitarian society that would be self-governing, for a moral discipline established, as it were, in the individual conscience. The citizens must be subject, within themselves, to a discipline which is not imposed merely by the fear of punishment. Now, the faith which will create this moral discipline better than any other is, according to Tocqueville (who on this point was still the disciple of Montesquieu), religious faith.

Besides this fundamental fact, American citizens are well informed; they are familiar with local affairs; they all have the advantage of a civic education. Finally, Tocqueville emphasized the role played by American administrative decentralization as opposed to French administrative centralization. American citizens acquire the habit of regulating their collective affairs at the level of the local district. As a consequence, they serve their apprenticeship in self-government in the limited sphere which they are in a position to know personally, and they extend the same spirit to affairs of state.

This theory of American democracy obviously differs from Montesquieu's theory, which was based on the ancient republics. But Tocqueville himself regarded his theory of modern democratic societies as the expansion and revision of the theory of the ancient republics developed by Montesquieu. I have come across a pertinent passage which does not appear in any of Tocqueville's books but occurs in the preparatory notes for Volume II of *Democracy in America*. In it, he compared his own interpretation of American democracy with Montesquieu's theory of the republic. Here is the passage.

> We must not take Montesquieu's idea in a narrow sense. What this great man meant is that the republic could continue to exist only through the influence of society upon itself. What he understands by virtue is the moral power which each individual exercises over himself and which prevents him from violating the right of others. When this triumph of man over temptation results from the weakness of the temptation or the consideration of personal interest, it does not constitute virtue in the eyes of the moralist, but it does enter into Montesquieu's conception, for he was speaking of the effect much more than the cause. In America it is not virtue that is great, but temptation that is small, which comes to the same thing. It is not disinterestedness that is great, it is interest that is taken for granted, which again comes to almost the same thing. Thus Montesquieu was right, although he spoke of ancient virtue, and what he said of the Greeks and Romans still applies to the Americans.

This passage enables one to present a sort of synthesis of the relation between Tocqueville's theory of modern democracy and Montesquieu's theory of the ancient republic.

There are, to be sure, essential differences between the republic as seen by Montesquieu and the democracy as seen by Tocqueville. The typical ancient democracy was equalitarian and frugal; the citizens tended to equality, but by a refusal to grant supremacy to commercial considerations. The ancient republics were equalitarian, but virtuous; they were equalitarian, but warlike. Modern democracy as seen by Tocqueville is fundamentally a commercial and industrial society. Therefore, it is impossible for self-interest not to be the dominant feeling in modern societies, and indeed it is on self-interest that modern democracy is necessarily founded. Thus it may be said that the principle, in Montesquieu's sense, of modern democracy, according to Tocqueville, is interest, not virtue. But, as the context indicates, interest and virtue have elements in common. In both cases, the citizens must be subject in themselves to a moral discipline. In both cases, the state can survive only through the influence which society itself exerts on its members. In both cases, the stability of the state is based on the discipline of its citizens and on the predominant influence which manners and beliefs exert on the conduct of individuals.

Such, broadly summarized, is the theory of American liberal democracy outlined by Tocqueville.

Let me repeat: he was by no means an entirely enthusiastic or uncritical admirer of American society or democracy. Tocqueville, who, at bottom, retained a hierarchy of values borrowed from the class to which he belonged or from which he was descended, i.e., the French aristocracy, was sensitive to the mediocrity which characterizes a civilization of this kind. He brought to modern democracy neither the enthusiasm of those who expected from it a transfiguration of the human lot nor the hostility of those who saw in it no less than the very decomposition of human society. Democracy, for him, was justified by the fact that it strove for the well-being of the greatest number; but this well-being would be without brilliance or grandeur, and it would always be attended by risks. Chief among these appear to be the following.

Every democracy tends toward centralization and therefore toward a kind of despotism which is capable of degenerating into tyranny. Democracy is in perennial danger of a tyranny of the majority. Every democratic regime postulates that the majority is right, and it is sometimes difficult to prevent a majority from abusing its victory and oppressing the minority.

Democracy, Tocqueville continued, tends to propagate the spirit of flattery, though it is understood that the sovereign whom the candidates for office will flatter is the people and not the monarch. But to flatter the popular sovereign is no better than to flatter the monarchical sovereign. It may even be worse, since the spirit of flattery in a democracy is what in ordinary language is called demagogy.

Tocqueville was very much aware of the two great racial problems confronting American society, which concerned relations between the whites and the Indians on the one hand and the whites and the Negroes on the other. If there was one problem that threatened the existence of the Union, it was unquestionably that of the Southern slaves. Tocqueville was darkly pessimistic, for he wrote to the effect that in proportion as slavery disappeared and legal equality tended to become established between Negroes and whites, the barriers that result from manners would be raised between the two races. In the last analysis, he said, there are only two solutions: either a pure and simple mingling of races, or segregation. But this mingling of races, said he, will not be accepted by the white majority. Therefore, segregation, after the end of slavery, will be almost inevitable. And he foresaw terrible conflicts between the two races.

As for relations between whites and Indians, I should like to quote a passage which is in Tocqueville's best style, so that one may hear the voice of this solitary man.

> The Spaniards pursued the Indians with bloodhounds, like wild beasts; they sacked the New World like a city taken by storm, with no discernment or compassion; but destruction must cease at last and frenzy has a limit: the remnant of the Indian population which had escaped the massacre mixed with its conquerors and adopted in the end their religion and their manners. The conduct of the Americans of the United States towards the aborigines is characterized, on the other hand, by a singular attachment to the formalities of law. Provided that the Indians retain their barbarous condition, the Americans take no part in their affairs; they treat them as independent nations

and do not possess themselves of their hunting-grounds without a treaty of purchase; and if an Indian nation happens to be so encroached upon as to be unable to subsist upon their territory, they kindly take them by the hand and transport them to a grave far from the land of their fathers.

The Spaniards were unable to exterminate the Indian race by those unparalleled atrocities which brand them with indelible shame, nor did they succeed even in wholly depriving it of its right; but the Americans of the United States have accomplished this twofold purpose with singular felicity, tranquilly, legally, philanthropically, without shedding blood, and without violating a single great principle of morality in the eyes of the world. It is impossible to destroy men with more respect for the laws of humanity.

Tocqueville did not respect the rule of modern sociologists, which is to abstain from value judgments and to refrain from irony. I should add that Tocqueville is probably wrong: the differences between American-Indian and Spanish-Indian relations are not just a matter of the attitudes adopted by each but also of the different densities of the Indian population in the north and the south. However this may be, this passage suggests once again the humanitarianism of the aristocrat. In France, we are often accustomed to think that only men of the left are humanitarian. But Tocqueville himself would have said that in France the radicals, the extreme republicans, were not humanitarians but revolutionaries, drunk with ideology and ready to sacrifice millions of men to their ideas. Tocqueville, as an aristocrat and a humanitarian, condemned the ideologists of the left, representatives of the French intellectual party, but also condemned the reactionary spirit of aristocrats who were nostalgic for an order that had vanished forever.

Tocqueville, therefore, was a sociologist who never ceased to judge while he described. In this sense, he belonged to the tradition of classical political philosophers who would not have conceived of analyzing regimes without judging them at the same time.

If I have emphasized the links between Tocqueville and Montesquieu, I have also suggested the links between Montesquieu and Aristotle. In the history of sociology, Tocqueville remains closest to classical philosophy, as interpreted by Prof. Leo Strauss.[6]

In Aristotle's eyes, one cannot interpret tyranny correctly unless one sees it as the regime furthest removed from the best of regimes, for the reality of the fact is inseparable from the quality of this fact. To try to describe institutions without judging them is to miss what constitutes them as such. Tocqueville remained in this tradition. His description of the United States was also an explanation of the means by which freedom is safeguarded in a democratic society, and it reveals throughout what threatens the equilibrium of American society. This very language implies a judgment, and Tocqueville did not believe he was violating the rules of social science by judging in and by his description. If he had been questioned, he would probably have replied (perhaps like Montesquieu, or in any case like Aristotle), that a description cannot be faithful unless it includes those judgments intrinsically related to the description, since in fact a regime is what it is by its own quality, and a tyranny can only be described as a tyranny.

In *Democracy in America,* Tocqueville was a sociologist in the style of Montesquieu or, as we might say, in the two styles Montesquieu bequeathed to us. A synthesis of the different aspects of a society is achieved in *The Spirit of the Laws* by means of the idea of the spirit of a nation. Thus, according to Montesquieu, sociology's prime objective is to comprehend the whole of a society. Tocqueville was certainly trying to comprehend the spirit of a nation in America, and to this end he employed the different categories which Montesquieu distinguished in *The Spirit of the Laws.* He established a distinction between historical causes and contemporary causes, geographical environment and historical tradition, the influence of laws and the influence of manners. These elements, taken as a whole, are regrouped in order to define in its singularity a unique society —American society. The description of this singular society results from the combining of different types of explanation of a more or less high degree of abstraction or generality.

But Tocqueville, in *Democracy in America,* aspired to a second sociological objective or, if you will, practiced another method. He raised a more abstract problem on a higher level of generality: the problem of democracy in modern societies. In this case we are dealing, so to speak, with an ideal type, comparable to the type of political regime which Montesquieu studied in the first part of *The Spirit of the Laws.* Turning to the abstract notion of a democratic society, Tocqueville asked what political form this democratic society can assume and why it assumes a certain form here and another elsewhere. In other words, starting with an ideal type—democratic society—he tried, by the comparative method, to reveal the influence of various causes by proceeding, to use his idiom, from the most general to the most particular.

Thus, there are two sociological methods in Tocqueville, of which one results in a portrait of a particular collectivity and the other raises the abstract historical problem of a society of a certain type.

III

I SHALL DEVOTE this section to an analysis of Alexis de Tocqueville's second major book, *The Old Regime and the French Revolution.* It goes without saying that I do not intend to give a detailed account of everything Tocqueville wrote about the France of the old regime and the Revolution but only to bring out the broad lines both of his method and of his interpretation.

The Old Regime and the French Revolution represents

an attempt comparable to Montesquieu's in his *Considerations on the Causes of the Grandeur and Decadence of the Romans*—an attempt at a sociological explanation of historical events. Because of the nature of this book, we necessarily discover the limits of sociological explanation—limits which Alexis de Tocqueville perceived just as clearly as did Montesquieu. Both believed that great events are explained by great causes but that the details of events are simply events, that is, they are not deducible from the structural facts of the society in question.

Up to a certain point, Tocqueville was studying France with America in mind. He was trying to understand why France had so much difficulty being a politically free society, although she was, or seemed to be, democratic, just as in the case of America he was looking for the causes of the reverse phenomenon, that is, the persistence of political liberty because or in spite of the democratic character of the society.

Hence, *The Old Regime and the French Revolution* is a sociological interpretation of a historical crisis, whose aim is to make events intelligible. At the outset, Tocqueville observed and reasoned in the manner of a sociologist. He refused to concede that the revolutionary crisis might be an accident pure and simple. He asserted that the institutions of the old regime themselves collapsed at the very moment that they were swept away by the revolutionary tempest. He went on to say that the revolutionary crisis had certain specific characteristics because it developed in the manner of a religious revolution.

The French Revolution's approach to the problems of man's existence here on earth was exactly similar to that of the religious revolutions as regards his afterlife. It viewed the "citizen" from an abstract angle, that is to say, as an entity independent of any particular social order, just as religions view the individual without regard to nationality or the age he lives in. It did not aim merely at defining the rights of the French citizen, but sought also to determine the rights and duties of men in general towards each other and as members of a body politic.

It was because the Revolution always harked back to universal, not particular, values and to what was the most "natural" form of government and the most "natural" social system that it had so wide an appeal and could be imitated in so many places simultaneously.

This coincidence of a political crisis with a kind of religious revolution is apparently characteristic of the great revolutions of modern societies. I need scarcely point out that the other great revolution of modern times, the Russian Revolution of 1917, had also, in the eyes of a sociologist of Tocqueville's school, the same quality of being religious in essence. I believe we may generalize the statement and say that every political revolution assumes some of the characteristics of a religious revolution when it aspires to universal validity, when it claims to be the way of salvation for all humanity.

Why, when all the institutions of the old regime were collapsing in Europe, was it in France that the revolution occurred? To this question Tocqueville gave an answer which is also sociological. But first I should like to quote a statement of Tocqueville's which is illuminating with regard to his method: "I speak of classes; these alone should concern the historian." This is verbatim. I am sure that if a magazine published this statement anonymously, four out of five readers would attribute it to Karl Marx. However the statement is not Marx's, it is Tocqueville's; and it follows the declaration, "I do not speak of individuals, who do not affect the essential phenomena of history, I speak of classes, which alone should concern the historian."

Those classes whose decisive roles Tocqueville mentioned are the nobility, the bourgeoisie, the peasantry, and secondarily, if you will, the workers. The classes he distinguished are, therefore, intermediate between the orders and estates of the old regime and the classes of modern societies. However, Tocqueville did not construct an abstract or sociological theory of classes. He did not define them or enumerate their characteristics; but he used the principal social groups of the old regime in France at the time of the Revolution to explain events.

What, then, are the principal phenomena which account for the fact of the collapse of the society of the old regime in France? The first phenomenon, as we well know, was administrative centralization and administrative uniformity. It is true that under the old regime France was characterized by an extraordinary diversity of legislation and regulation from one region to another, as Tocqueville told us; but the fact is that royal administration, administration by intendants, was gradually becoming the effective force. Diversity was merely a meaningless survival; France was centrally and uniformly administrated well before the Revolution broke out.

Here is the most characteristic passage.

> Some have been surprised at the extraordinary ease with which the Constituent Assembly annihilated at one fell swoop all the historic divisions of France—the provinces—and split up the kingdom into eighty-three well-defined units—the departments—almost as if they were partitioning not an ancient kingdom but the virgin soil of the New World. Nothing surprised and, indeed, shocked the rest of Europe more than this. "It is the first time," Burke said, "that we have seen men hack their native land to pieces in so barbarous a manner." But though it might seem that living bodies were being mutilated, actually the "victims" were already corpses.

Here are a few more words on Paris and administrative centralization.

> During the same period as that in which Paris was coming to dominate the entire country, another change was taking place within the city itself, a change which all historians do well to take into account. Besides being at once a business and commercial center, a city of pleasure seekers and consumers, Paris had now de-

veloped into a manufacturing city—a change which, in conjunction with the political ascendancy of Paris described above, was destined to have great and dangerous consequences.

In other words, according to Tocqueville, Paris had become an industrial center, the industrial center of France, even before the end of the eighteenth century: "Though the statistical records of the old regime are far from trustworthy, we are justified, I think, in saying that during the sixty years preceding the Revolution the number of workers employed in Paris more than doubled, while in the same period the general population of the city rose by hardly a third."[7]

Secondly, in this France which was centrally administrated and in which, increasingly, the same regulations applied throughout the entire territory, society had crumbled into particles, as it were. The French were not able to discuss their affairs, because the essential condition for the formation of the body politic—namely, freedom—was lacking.

Tocqueville gives a purely sociological description of what Durkheim would have called the disintegration of French society. There was no longer any unity among the privileged classes or among the different classes of the nation, for two reasons: The first, as I have just indicated, was the lack of political liberty; and the second was the separation between the privileged groups of the past, which had lost their historical function and retained their privileges, and the groups of the new society, which were playing a decisive role but remained separate from the old nobility.

No doubt it was still possible at the close of the eighteenth century to detect shades of differences in the behavior of the aristocracy and that of the bourgeoisie; for nothing takes longer to acquire than the surface polish which is called good manners. . . . But basically all who ranked above the common herd were of a muchness; they had the same ideas, the same habits, the same tastes, the same kinds of amusements; read the same books and spoke in the same way. They differed only in their rights.

I doubt if this leveling-up process was carried so far in any other country, even in England, where the different classes, though solidly allied by common interests, still differed in mentality and manners. For political freedom, though it has the admirable effect of creating reciprocal ties and a feeling of solidarity between all the members of a nation, does not necessarily make them resemble each other. It is only government by a single man that in the long run irons out diversities and makes each member of a nation indifferent to his neighbor's lot.

Here, I think, is the center of Tocqueville's sociological analysis. The different privileged groups of the French nation were moving simultaneously toward uniformity and toward separation; they were like each other in reality, but were separated by privileges, manners, and traditions, and in the absence of political liberty they did not succeed in acquiring that sense of solidarity indispensable to the health of the body politic.

The segregation of classes, which was the crime of the late monarchy, became at a late state a justification for it, since when the wealthy and enlightened elements of the population were no longer able to act in concert and to take part in the government, the country became, to all intents and purposes, incapable of administrating itself and it was needful that a master should step in.

Consider this passage for a moment. What is immediately impressive is the more or less aristocratic conception of the government of societies, characteristic of both Montesquieu and Tocqueville. For both writers, the government of a nation can be exercised only by the rich and enlightened element, and neither writer hesitated to juxtapose the two adjectives *rich* and *enlightened*. They were not demagogues, and the relation between the two terms seemed obvious to them. They were writing in an age when those who possessed no material means did very rarely have the chance to become educated and consequently were not enlightened. Do not think them cynical; to them it was self-evident; and it was true that in the eighteenth century only the wealthy element of the nation could be enlightened.

Secondly, Tocqueville thought he saw—and I think he saw correctly—that in France the characteristic phenomenon at the origin of the Revolution, and, I should add, of all French revolutions to follow, is the inability of the privileged groups to reach an agreement on a mode of government for their country. This phenomenon explains that frequency of changes of regime which persists even today. Tocqueville would add that administrative centralization increases throughout the nineteenth century and even into the twentieth, and the rich and enlightened element of the nation continues to be unable to agree, with the result that from time to time democratic government becomes impossible and a leader must intervene.

This analysis of French politics is, I believe, exceptionally lucid. It may be applied to the whole political history of France in the nineteenth and twentieth centuries. It accounts for the curious phenomenon that, of the countries of Western Europe, France has been, in the nineteenth and until a recent date in the twentieth century, the one that has changed the least economically and socially and also the one that has been perhaps the most turbulent politically. This combination of economic and social semi-conservatism and political turbulence is, I think, readily explained in the context of Tocqueville's sociology. The phenomenon is much harder to explain if we try to find a term-by-term correspondence between social and political data.

Another quotation, to conclude this analysis:

Yet, when sixty years ago the various classes which under the old order had been isolated units in the social system came once again in touch, it was on their sore spots that they made contact and their first gesture was to fly at each other's throats. Indeed, even today [i.e., a hundred years ago], though class distinctions are no more, the jealousies and antipathies they caused have not died out.

The central theme of Tocqueville's interpretation of French society is, therefore, that at the end of the old regime, France was, of all European societies, at once the most democratic, in Tocqueville's sense of the word—i.e., the one in which the tendency to uniformity of conditions and social equality of persons and groups was most marked —and also the one in which political freedom, freedom of political discussion, was most curtailed and society most crystallized in traditional institutions which corresponded less and less to the reality of France.

In other words, the Tocquevillian concept of revolution is essentially political. It is the resistance of the political institutions of the past to the modern democratic movement which risks provoking, in one place or another, an explosion. Tocqueville also pointed out that this kind of revolution will occur not when things are going worse, but when they are going better. There is a chapter in *The Old Regime and the French Revolution* in which he showed that things had never gone so well under the old regime as during the reign of Louis XVI.[8]

I need scarcely emphasize that Tocqueville would not doubt for a single moment that the Russian Revolution fits into his political schema of revolutions much better than into the Marxist one. The Russian economy entered a growth cycle in the 1880's; between 1880 and 1914 the rate of growth of the Russian economy was one of the highest, and perhaps the highest, in Europe.[9] Furthermore, the Russian Revolution began with a revolt against the political institutions of the old regime, in the sense in which we speak of the old regime apropos of the French Revolution. If the party that seized power in Russia claimed to represent a completely different ideology, Tocqueville would have had no trouble incorporating these phenomena into his system, for in his eyes it was characteristic of democratic revolutions to act in the name of freedom and to tend toward political and above all administrative centralization. He referred several times to the possibility of a state which would attempt to manage the economy as a whole.

Therefore, from the point of view of Tocqueville's theory, the Russian Revolution is the collapse of the political institutions of the old regime in a general period of society's modernization. An explosion favored by the extension of a war, the Revolution resulted in a government which claimed to represent the democratic ideal but which pushed to its conclusion the idea of administrative centralization and of state management of society as a whole.

If we ask what attitude Tocqueville adopted toward those two alternatives which have obsessed historians of the French Revolution, namely, catastrophe or blessing, necessity or accident, I think his answer would be that he would refuse to subscribe to either of the extreme positions. Obviously the French Revolution was not, in his eyes, an accident pure and simple. It was necessary, if by this we understand that the democratic movement was bound eventually to sweep away the institutions of the old regime; but it was not necessary in the precise form it assumed and in the details of its episodes. Was it beneficial or catastrophic? Tocqueville's answer would probably be: both. Or more precisely, his book contains all the elements of the criticism advanced by men of the right with respect to the French Revolution, but it also contains the justification of what happened, with a sort of nostalgia that things did not happen otherwise.

The elements of criticism of the French Revolution have to do first with those men of letters known in the eighteenth century as *philosophes* and in the twentieth century as intellectuals. As is well known, philosophers, men of letters, and intellectuals frequently criticize one another. Tocqueville commented on the role played by writers in eighteenth-

century France and in the Revolution, just as we continue
to comment, with admiration or regret, on the role they
play today. Here is the passage I find most characteristic.

> Our men of letters did not merely impart their revo-
> lutionary ideas to the French nation; they also shaped
> the national temperament and outlook on life. In the
> long process of molding men's minds to their ideal
> pattern their task was all the easier since the French
> had had no training in the field of politics, and they
> thus had a clear field. The result was that our writers
> ended up by giving the Frenchman the instincts, the
> turn of mind, the tastes, and even the eccentricities
> characteristic of the literary man. And when the time
> came for action, these literary propensities were im-
> ported into the political arena.
>
> When we closely study the French Revolution we
> find that it was conducted in precisely the same spirit
> as that which gave rise to so many books expounding
> theories of government in the abstract. Our revolution-
> aries had the same fondness for broad generalizations,
> cut-and-dried legislative systems, and a pedantic sym-
> metry; the same contempt for hard facts; the same
> taste for reshaping institutions on novel, ingenious,
> original lines; the same desire to reconstruct the entire
> constitution according to the rules of logic and a pre-
> conceived system instead of trying to rectify its faulty
> parts.

And the conclusion, whose equivalent may be found in
some of today's newspapers:

> The result was nothing short of disastrous; for what
> is a merit in the writer may well be a vice in the states-
> man and the very qualities which go to make great
> literature can lead to catastrophic revolutions.

A passage of this kind is the basis for a whole literature.
For example, Taine's first book, *Les Origines de la France
Contemporaine,* is scarcely more than a development of
this theme of the role of writers and men of letters.[10] To
correct, one might say, what is excessive about this inter-

pretation, I recommend an excellent book[11] by Daniel Mornet on the intellectual origins of the French Revolution, in which he endeavors to demonstrate that to a very large extent the writers and men of letters did not resemble the image of them provided by Tocqueville or Taine.

Also among the elements of criticism is an analysis of what Tocqueville called the fundamental irreligion which had spread within one part of the French nation. We have noted the mingling of the spirit of religion and the spirit of liberty which according to Tocqueville is the foundation of American liberal democracy. We read in *The Old Regime and the French Revolution* of the opposite situation;[12] that is, the element of the country which had become ideologically democratic had not only lost its faith but had become anticlerical and antireligious. Moreover, Tocqueville declared himself full of admiration for the clergy of the old regime,[13] and he boldly and explicitly expressed his regret that it had not been possible to preserve, at least partially, the role of the aristocracy in modern society.

And precisely because this idea is not a fashionable one and because it is altogether characteristic of Tocqueville, I shall quote still another passage.

> When we read the *cahiers* they presented to the Estates-General, we cannot but appreciate the spirit and some of the high qualities of our aristocracy, despite its prejudices and failings. It is indeed deplorable that instead of being forced to bow to the rule of law, the French nobility was uprooted and laid low, since thereby the nation was deprived of a vital part of its substance, and a wound that time will never heal was inflicted on our national freedom. When a class has taken the lead in public affairs for centuries, it develops as a result of this long, unchallenged habit of preeminence a certain proper pride and confidence in its strength, leading it to be the point of maximum resistance in the social organism. And it not only has itself the manly virtues; by dint of its example it quickens them in other classes. When such an element of the body politic is forcibly excised, even those most

273

hostile to it suffer a diminution of strength. Nothing can ever replace it completely, it can never come to life again; a deposed ruling class may recover its titles and possessions but nevermore the spirit of its forebears.

Bernanos has written pages which, without Tocqueville's precision of analysis, reach the same conclusion. It is not enough to have the institutions of freedom: elections, parties, a parliament. Men must also have a certain taste for independence, a certain sense of resistance to power, for freedom to be authentic.

The judgment which Tocqueville passed on the Revolution, the feelings which inspired him, are precisely the ones Auguste Comte would have declared aberrant. In Comte's eyes, the attempt of the Constituent Assembly was doomed because it aspired to a synthesis of the Catholic, theological, feudal institutions of the old regime and the institutions of modern times. But, said Comte with his customary intransigence, a synthesis of institutions based on radically different attitudes is impossible. Tocqueville would have preferred, not that the democratic movement should not sweep away the institutions of the old France, for that movement was irresistible and inevitable, but that as many institutions of the old regime as possible be preserved to contribute to the safeguarding of freedom in a society dedicated to the pursuit of well-being and doomed, as it were, to social revolution.

Thus, for a sociologist like Comte, the synthesis of the Constituent Assembly was impossible from the outset. For a sociologist like Tocqueville, this synthesis—whether possible or not, he did not decide—would in any case have been desirable. Politically we might say that Tocqueville was in favor of the first French Revolution, that of the Constituent Assembly; and it is to the Constituent Assembly that his mind returned, nostalgically as it were. For him, the great moment of the French Revolution, the great moment of France, was 1788 and 1789, when the French people were inspired by a boundless confidence and hope.

Here is a passage characteristic of this aspect of Tocqueville's thinking.

I think that no epoch of history has ever witnessed so large a number so passionately devoted to the public good, so honestly forgetful of themselves, so absorbed in the contemplation of the common interest, so resolved to risk everything they cherished in their private lives, so willing to overcome the small sentiments of their hearts. This was the general source of that passion, courage, and patriotism from which all the great deeds of the French Revolution were to issue.

The spectacle was short, but it was one of incomparable grandeur. It will never be effaced from the memory of mankind. All foreign nations witnessed it, applauded it, were moved by it. There was no corner of Europe so distant and secluded that this glow of admiration and of hope did not reach it. In that immense mass of memoirs left to us by the contemporaries of the Revolution, I have found none in which the recollection of those first days of 1789 has not left imperishable traces; everywhere they reveal the clarity, the freshness, the vivacity of the impressions of youth.

I venture to say that there is but one people on this earth which could have staged such a spectacle. I know my nation—I know but too well her errors, her faults, her foibles, and her sins. But I also know of what she is capable. There are enterprises which only the French nation can conceive; there are magnanimous resolutions which this nation alone dares to take. She alone will suddenly embrace the common cause of humanity, willing to fight for it; and if she be subject to awful reverses, she has also sublime moments which sweep her to heights which no other people will ever reach.

We see how Tocqueville, who is considered—and indeed was—a critic of France, who compared France's evolution to that of the Anglo-Saxon countries with a sort of regret that the political history of France had not been that of England or the United States, was at the same time capable of turning self-criticism into self-glorification, for that passage "France alone. . . ." might evoke further periods on the unique vocation of the fatherland. Tocqueville tried to make events sociologically intelligible; but there was in the background, with him as with Montesquieu, the idea of national character, or national history. A few observations on this point might be useful.

Earlier I mentioned Tocqueville's criticism of the role of men of letters. In this chapter on men of letters, he refused to accept the explanation for their activity in terms of national character. On the contrary, he said that the role played by men of letters had nothing to do with the spirit of the French nation and was explained by social conditions. It was because there was no political freedom, because men of letters did not participate in public affairs, because they were ignorant of the real problems of government, that they became lost in abstract theories. Tocqueville's chapter on men of letters is the first example of an analysis, very fashionable today, of the role of intellectuals in a society in the process of modernization, when these men of letters have no experience with the problems of government and are drunk with ideology.

On the other hand, when he was discussing the French Revolution and its period of greatness, Tocqueville tended to draw a portrait—I shall not bother to quote the celebrated description at the end of *The Old Regime and the French Revolution*—a sort of synthetic portrait, in the style of Montesquieu, of the spirit of a nation. By a synthetic portrait I mean a description of a collectivity's mode of conduct, without this mode of conduct being offered as a final explanation. This mode of conduct is as much a result as a cause, but it is sufficiently original, sufficiently specific for the sociologist, after an analysis of its different aspects, to marshal the different features into an over-all portrait.[14]

The second volume of *The Old Regime and the French Revolution* was to have discussed the sequence of events and the role of men, accidents, and chance. In the notes which have been published there are many indications as to the role of individual men. I shall quote only a few such lines.

I am less struck by the genius of those who made the Revolution because they desired it than by the singular imbecility of those who made it without desiring it. When I consider the French Revolution, I am amazed at the prodigious magnitude of the event, at the glare it cast to the extremities of the earth, at its power, which more or less stirred every nation.

When I, then, turn to that court which had so great a share in the Revolution, I see there some of the most trivial scenes in history: harebrained or narrow-minded ministers, dissolute priests, futile women, rash or mercenary courtiers, a King with peculiarly useless virtues.

The contrast this passage affords has more than a literary value. It represents, I think, the over-all vision which Tocqueville would have given us had he been able to finish his book. After playing the role of sociologist in his study of origins, after showing how postrevolutionary society was to a large extent prepared by prerevolutionary society in its stress on uniformity and administrative centralization, he would then have tried to follow the course of events and would not have omitted what, for Montesquieu as for himself, was the essence of history, namely, the event at a given juncture—a series of contingent circumstances or a decision made by one man, all of which might easily be imagined otherwise. If, at a certain date, Louis XVI had made a certain other decision, the course of events would have been different. There is, in Tocqueville, one level on which the necessity of the historical movement appears and another level on which we rediscover the role of men.

The essential fact is the failure of the Constituent Assembly, that is, the failure of a synthesis between the virtues of the aristocracy or monarchy and the democratic movement. It is the failure of this synthesis which in Tocqueville's eyes accounted for France's difficulty in finding a political equilibrium. Tocqueville believed that the France of his time—nineteenth-century France—required a monarchy; at the same time he saw clearly the weaknesses of a monarchy. He believed that political liberty cannot be stabilized unless a limit is placed on administrative centralization and uniformity; but this centralization and this administrative despotism seemed to him part of the democratic movement. The same kind of analysis which explains the liberal vocation of American democracy explains the antiliberal bias in democratic France.

Here is one last remark, typical of the political attitude of men like Tocqueville, men of the center, and their criticism of extremes: "In brief," he wrote, "I can conceive up to now how a man of education, common sense and good intentions might become a radical in England. I have never been able to conceive of the reconciliation of these three things in the French radical."

A similar witticism about the Nazis was popular thirty years ago: The Germans were all intelligent, honorable, and pro-Hitler, but they never had more than two of these qualities at once. Tocqueville was saying that a man of education, common sense, and good intentions could not be a radical in France. If he were educated and had good intentions, he did not have common sense. If he were educated and had common sense, he did not have good intentions. In Tocqueville's eyes, the French revolutionary often had good intentions and great abstract intelligence, but he lacked common sense. It goes without saying that in politics common sense is very largely a matter of personal preference; Auguste Comte would not have hesitated to declare Tocqueville's nostalgia for the synthesis of the Constituent Assembly devoid of common sense. Perhaps this justifies a sociological formula on the "relativity of common sense."

IV

I HAVE, up to now, presented two aspects of Alexis de Tocqueville's sociological method: first, the portrait cf a particular society, American society; and next, the sociological interpretation of a historical crisis, the French Revolution. I turn now to an examination of the second volume of *Democracy in America,* in which I find exemplified a third method which is characteristic of the author: the creation of a kind of ideal type of democratic society from which he deduced some of the tendencies of future society.

The second volume of *Democracy in America* differs from the first in the method used and the questions raised. It is almost an example of what might be called an intellectual experiment. Tocqueville established a priori the structural traits of a democratic society which, as we know, is characterized by a gradual disappearance of distinctions of class or rank and an increasing uniformity of living conditions. Having established these structural traits of democratic society, Tocqueville then raised these four questions: What are the results for intellectual activity? for the feelings of the Americans? for manners properly so called? and, finally, for political society?

The undertaking is in itself difficult, and even audacious, for two main reasons:

(1) It is not certain that, given the structural traits of a democratic society, one can determine the state of intellectual activity or of manners. In the jargon of modern sociology, this problem belongs to the sociology of knowledge. To what extent does social context determine the form taken by various intellectual activities? This sociology of knowledge has an abstract and therefore dangerous quality. It may turn out that the prose, poetry, drama, and parliamentary eloquence of the various democratic societies will be just as heterogenous in the future as these intellectual activities have been in the past.

(2) As for the structural traits of democratic society which Tocqueville took as his point of departure, some are related to the peculiarities of American society, and others are inseparable from the essence of democratic society. This ambiguity involves another as to the degree of generality of the answers Tocqueville can give to the question he has raised.[15] The answers to the questions raised in the second volume will sometimes be on the order of a *tendency*. Sometimes the answers will be on the order of an *alternative*. Or it may turn out that there is no possible answer to a question stated in such general terms.

Having expressed these initial reservations, I shall now review the four principal parts of the book, to give an idea of Tocqueville's answers.

Opinions of the second volume vary considerably. As soon as the book was published, there were critics who denied the author the favor they had bestowed on the first. It may be said that in this second volume Tocqueville outdid himself, in all senses of the phrase. He was more Tocquevillian than ever; he exhibited a great capacity for reconstruction or deduction from a small number of facts—a capacity which sociologists sometimes admire and historians are more inclined to question.

In the first part of the book—the one devoted to setting forth the consequences of democratic society for intellectual activity—Tocqueville took up in turn the attitude toward ideas, certain specific traits of religions in a democratic society, and the different literary genres: poetry, drama, oratory. What were his leading ideas on the subject?

The first idea I should like to point out, which is of a very general character, recalls one of Tocqueville's favorite comparisons between the French and the Americans. In Chapter IV, "Why the Americans Have Never Been so Eager as the French for General Ideas in Political Affairs," the answer he gives is as follows.

The Americans are a democratic people who have always directed public affairs themselves. The French are a democratic people who for a long time could only speculate on the best manner of conducting them. The social condition of the French led them to conceive very general ideas on the subject of government, while their political constitution prevented them from correcting those ideas by experiment and from gradually detecting their insufficiency; whereas in America the two things constantly balance and correct each other.

This explanation belongs to the sociology of knowledge, but it is of a simple and empirical type. The essential interpretation of the contrast between the empirical approach of the Americans and the ideological approach of the French is that the French have acquired a taste for ideology, because for centuries they have not been able actually to participate in public affairs. The less one participates in public affairs, the more one develops theories.

A second idea, which, I think, goes further and is also more speculative but not without interest, is the interpretation of certain religious beliefs in terms of the society. The example I am taking is from Chapter V, which discusses the relation between the democratic instincts and the mode of religious belief. Here is the most typical passage.

> The preceding observation, that equality leads men to very general and very vast ideas, is principally to be understood in respect to religion. Men who are similar and equal in the world readily conceive the idea of the one God, governing every man by the same laws and granting to every man future happiness on the same conditions. The idea of the unity of mankind constantly leads them back to the idea of the unity of the Creator; while on the contrary, in a state of society where men are broken up into very unequal ranks, they are apt to devise as many deities as there are nations, castes, classes, or families, and to trace a thousand private roads to heaven.

Here we see another modality of an interpretation which

belongs to the sociology of knowledge. This is the growing uniformity of increasing numbers of individuals, not integrated into separate groups, which induces them to conceive both the unity of mankind and the unity of the Creator.

Explanations of this sort are also to be found in Comte. The one I have just presented is certainly much too simple; it is this kind of generalizing analysis that has prejudiced many historians against Tocqueville.

Here is another example I should like to mention. Tocqueville suggested that a democratic society tends to believe in the unlimited perfectibility of human nature. In democratic societies, social mobility reigns and each individual has the hope or expectation of rising in the social hierarchy. A society in which ascent is possible tends to conceive on the metaphysical level a comparable ascent for humanity as a whole. An aristocratic society, in which each man receives his rank at birth, is reluctant to believe in the unlimited perfectibility of humanity, because this belief would contradict the ideological formula on which that society rests. On the other hand, the idea of progress is almost consubstantial with a democratic society.[16]

In this case we see not only the transition from the social organization to a certain ideology, but also the intimate relation between social organization and ideology, the latter providing the foundation, as it were, for the former.

I should also like to mention the chapter in which Tocqueville explained that Americans are more apt to distinguish themselves in the applied sciences or in the application of science to industry than in the pure sciences. This has been true over a long period of time; it is almost a fact of experience. But Tocqueville, in his own style, showed that since a democratic society is essentially concerned with well-being, it is obliged not to devote the same concern to the pure sciences as a society of the more aristocratic type, in which those who devote themselves to the sciences are men of wealth, luxury, and leisure.[17]

A final example from this first part of Volume II is the

relation between democracy, aristocracy, and poetry.[18] I shall quote a few lines, not this time to make you admire Tocqueville but, on the contrary, to illustrate the flights of the abstract imagination.

> Aristocracy naturally leads the human mind to the contemplation of the past and fixes it there. Democracy, on the contrary, gives men a sort of instinctive distaste for what is ancient. In this respect aristocracy is far more favorable to poetry; for things commonly grow larger and more obscure as they are more remote, and for this twofold reason they are better suited to the delineation of the ideal.

Here we see that it is possible, by beginning with a small number of facts, to construct a theory which would be correct if there were only one kind of poetry, if poetry could flourish only by virtue of the idealization of beings and objects remote in time.

Regarding historians, Tocqueville shrewdly observed that democratic historians will tend to explain events in terms of anonymous forces, the irresistible mechanisms of historical necessity, while aristocratic historians will tend to place the emphasis on the role of great men.[10] Undoubtedly he was right. The theory of historical necessity—the theory that denies the power of accidents and of great men —undoubtedly belongs to the democratic age in which we live.

Let us proceed to the second part, in which Tocqueville tries, always in terms of the structural traits of democratic society, to elucidate the feelings which will be fundamental to every democratic society. Here, I think, summarized in a style slightly different from his own, are the conclusions at which he arrived.

In a democratic society, there will prevail a passion for equality which will necessarily triumph over the taste for liberty. A democratic society will be more concerned to erase inequalities between individuals and groups than to maintain respect for legality or personal independence. A democratic society will be motivated by a taste for material well-being; it will be tormented by a kind of perpetual restlessness, precisely because material well-being and equality cannot create a calm and contented society, since everyone compares himself with others and prosperity is never certain. Whence the quality of democratic societies, as seen by Tocqueville: restless, but fundamentally unchanging.

These superficially turbulent societies, as we may call them, will tend to be free, but it is to be feared that men love liberty as the condition of material well-being rather than for its own sake. And, Tocqueville added, it is conceivable, therefore, that under certain circumstances, when free institutions seem to be functioning badly and endangering prosperity, men will be inclined to sacrifice freedom in the hope of ensuring the well-being to which they aspire.

In a society of this kind, all professions will be considered honorable, because all professions will be fundamentally the same: they will all be wage earning. Democratic society, Tocqueville implied, is the society of the universal wage earner. But the society of the universal wage earner tends to do away with differences of kind and essence between the so-called noble and the so-called ignoble activities. The distinction between, say, domestic service and the free professions will gradually tend to disappear, each becoming in its own right *a job* (to use the modern idiom)— a job which yields a certain income, a certain salary. To be sure, there will continue to be inequalities of prestige from one occupation to another, according to the importance of the salary attached to each; but there will be no difference of kind.

A society of this kind, Tocqueville continued, is an individualistic society in which each person, with his family, tends to become isolated from the others. Curiously, an individualistic society of this kind reveals certain common traits with the isolation characteristic of despotic societies. Despotism, Tocqueville said, tends to isolate individuals from one another. Now this restless, materialistic society, dedicated to well-being, creates a comparable isolation between families, but it does not follow that this individualistic society is doomed to despotism, for certain institutions may prevent this transition. These institutions are the associations freely created by the initiative of individuals, associations which may and must be interposed between the solitary individual and the all-powerful state.

A society of this type risks public administration of all social activities. Democracy, as has been already observed, tends normally toward centralization. Tocqueville conceived of a society totally planned by the state. But from his point of view, this administration, which would embrace society as a whole and which in certain respects is realized in the society known today as socialist, far from constituting the ideal of the disalienated society to succeed capitalist society, represents the despotic society most to be feared.

A democratic society is on the whole materialistic, if by this we mean that it is concerned with acquiring as much worldly wealth as possible and with permitting the greatest possible number of individuals to live as well as possible. But, Tocqueville added, in contrast to the surrounding materialism, there occasionally occur outbursts of exalted spiritualism. A materialistic society of this kind is given, as it were, to eruptions of religious exaltation, sometimes of the most disembodied type. This eruptive spiritualism is contemporaneous with a normalized and habitual materialism. These two contrary phenomena both constitute the essence of a democratic society.

On these different points I have not yet quoted any texts, which are to be found almost anywhere. Here is one.

> Equality every day confers a number of small enjoyments on every man. The charms of equality are every instant felt and are within the reach of all; the noblest hearts are not insensible to them, and the most vulgar souls exult in them. The passion that equality creates must therefore be at once strong and general.

And again:

> I think that democratic communities have a natural

taste for freedom; left to themselves, they will seek it, cherish it, and view any privation of it with regret. But for equality their passion is ardent, insatiable, incessant, invincible; they call for equality in freedom; and if they cannot obtain that, they still call for equality in slavery. They will endure poverty, servitude, barbarism, but they will not endure aristocracy.

This passage is altogether typical of Tocqueville. It reveals two characteristics of his intellectual development: the attitude of the aristocrat of old family who is sensitive to the rejection by modern societies of the noble tradition, and also the intellectual context of his thought, the influence of Montesquieu, as evidenced in the dialectical play on the two concepts of freedom and equality. You remember that in Montesquieu's theory of political regimes, the essential dialectic is that of freedom and equality. The freedom of monarchies is based on the distinction of orders and the feeling of honor; the equality of despotism is an equality in bondage. Tocqueville adopted Montesquieu's dialectic and showed that the predominant feeling in democratic societies is the desire for equality at any price, which can lead to the acceptance of domination but does not necessarily imply slavery.

To conclude this section, here is still another passage in which Tocqueville gave the underlying reason why in a democratic society all the professions are ultimately of the same type: "No profession exists in which men do not work for money; and the remuneration that is common to them all gives them all an air of resemblance."

Here we have Tocqueville at his best: an apparently commonplace, very general fact, from which he draws a series of very far-reaching conclusions; for at the time he was writing, the tendency was making its first appearance; today it has been expanded and intensified. One of the least doubtful characteristics of American society is precisely this conviction that all the professions are honorable or, more accurately, are essentially of the same nature. Tocqueville continues:

This serves to explain the opinions that the Americans entertain with respect to different callings. In America no one is degraded because he works, for everyone about him works also; nor is anyone humiliated by the notion of receiving pay, for the President of the United States also works for pay. He is paid for commanding, other men for obeying orders. In the United States professions are more or less laborious, more or less profitable; but they are never either high or low: every honest calling is honorable.

To be sure, this is an ideal representation, and the picture could stand some shading. But it seems to me to be fundamentally accurate.

I shall now turn to the third part of Volume II, relating to manners, and I shall consider, above all, Tocqueville's ideas on the subject of revolutions and wars.

Here is the general picture Tocqueville drew: He began by explaining that, in democratic societies, manners have a tendency to grow less rigorous, that relations between Americans tend to be simple and easy, not very formal or stylized. The subtle and delicate refinements of aristocratic *politesse* tend to disappear in a sort of "good Joe-ism," to use a modern idiom. The style of interpersonal relations in the United States is direct, and this style tends to spread throughout American society as a whole. Relations between masters and servants tend to be of the same type as relations between people supposedly of good society. In other words, the gradations of aristocratic hierarchy which Tocqueville said survive in the interpersonal relations of European societies increasingly disappear in a fundamentally equalitarian society like the American one.

Tocqueville was aware that this phenomenon is related to the peculiarities of American society, but he was tempted to believe that European societies will evolve in the same direction, that this equalitarian style will gradually tend to spread in proportion as societies become more democratic, though it is apparent that societies with a long aristocratic past like the European ones will have more difficulty adopting this equalitarian style.

Here, then, is what Tocqueville said about wars and revolutions as a result of this ideal type of democratic society.

The first idea, which I have already formulated in discussing the French Revolution, is that great political or intellectual revolutions belong to a transitional phase between traditional and democratic societies and not to the essence of democratic societies. In other words, he said, in democratic societies great revolutions will become rare. Nevertheless, these societies will be essentially unsatisfied. On rereading Tocqueville, I noticed an idea which I had thought more or less my own, namely, the disputatious satisfaction of modern industrial societies.[20] Tocqueville wrote that democratic societies can never be satisfied because, being equalitarian, they are envious, but that in spite of this superficial turbulence, they are fundamentally conservative.

Why are democratic societies antirevolutionary? The underlying reason is that, as living conditions improve, the number of those who have something to lose by a revolution increases. Too many individuals and classes possess something in democratic societies for the latter to be ready to risk what they have on the gamble of revolution.[21] Tocqueville wrote:

> It is believed by some that modern society will be always changing its aspect; for myself, I fear that it will ultimately be too invariably fixed in the same institutions, the same prejudices, the same manners, so that mankind will be stopped and circumscribed; that the mind will swing backwards and forwards forever without begetting fresh ideas; that man will waste his strength in bootless and solitary trifling, and though in continual motion, that humanity will cease to advance.

Tocqueville was both right and wrong. He was perhaps right on the point that highly developed democratic societies are more disputatious than revolutionary. But he was wrong to underestimate the principle of the movement that carries along modern democratic societies: the development of science and industry. He had a tendency to combine two images, fundamentally stabilized societies and societies which were fundamentally preoccupied with well-being; but what he did not see clearly enough is that a preoccupation with well-being combined with a passion for science results in an uninterrupted series of scientific and technical discoveries. Apart from the fundamental social conservatism of democratic societies, a revolutionary principle is at work: science.

Tocqueville had been educated in a family whose memories of the great Revolution were tragic. His father had narrowly escaped being guillotined at the time of the Terror, and several of his relatives, notably Malesherbes, Louis XVI's lawyer, had been executed. Tocqueville was instinctively hostile to revolutions, and like all of us, he found convincing reasons to justify his feelings.[22] One of democratic society's best protections against despotism, he said, is a respect for legality. But revolutions, by definition, violate legality. They accustom people to not bowing before the law. This acquired contempt for the law survives the revolutions and, consequently, becomes a possible cause of despotism. Tocqueville was inclined to believe that the more frequently revolutions occurred in democratic societies, the greater the danger the society ran of becoming despotic.

This may be a justification of a predisposition; but it does not, therefore, follow that the justification is false. Let us say, with an allusion to contemporary events, that once a constitution has been violated, it is easier to violate it a second and third time. With constitutions, it is most of all the first step that counts. It was this exposure to illegality, this contempt for constitutional legality, which Tocqueville feared in revolutions.

As for war, Tocqueville thought that democratic societies would be little inclined to it. Further, democratic societies would be quite incapable of preparing for war in time of peace, but, once war had begun, they would be incapable of ending it. In other words, he had drawn a rather faithful portrait of the foreign policy of the United States until quite recently.

War is regarded by democracies as a disagreeable interlude in normal existence, which is peaceful. When democratic men are at peace, they think about war as little as possible and thus are ill prepared for it, which means that normally they suffer defeats in the first battles. If, Tocqueville continued, a democratic state is not utterly defeated in the course of the first battles, then eventually it becomes completely mobilized and wages war to its end in total victory.

Because democratic societies are little inclined to war does not mean that they will not wage war. Tocqueville feared—and on this point I believe he was mistaken—that in democratic societies the armies would be warmongers, as we would say nowadays. In a classic analysis, he showed that professional soldiers, who enjoy only a mediocre prestige in peace time, would encounter in the area of advancement the standard difficulties resulting from the low mortality rate of officers in normal times. Under these circumstances, professional soldiers—and particularly noncommissioned officers—would be more inclined to desire war than ordinary men. I confess I am rather uneasy about these detailed prophecies which are, of course, characteristic of an excessive propensity to generalization.

If despots appeared in democratic societies, they would be tempted to wage war, both to strengthen their power and to satisfy their armies. I refer you to Chapter XXIII, which is entitled, "Which Is the Most Warlike and Most Revolutionary Class in Democratic Armies."[23] Here, however, is a passage from Chapter XXIV which is not without interest.

> When a war has at length, by its long continuance, roused the whole community from their peaceful occupations and ruined their minor undertakings, the same passions that made them attach so much importance

to the maintenance of peace will be turned to arms. War, after it has destroyed all modes of speculation, becomes itself the great and sole speculation, to which all the ardent and ambitious desires that equality engenders are exclusively directed. Hence it is that the selfsame democratic nations that are so reluctant to engage in hostilities sometimes perform prodigious achievement when once they have taken the field.

This is a fine description of the total war of the democratic societies in the twentieth century.

Finally, we come to the fourth and last part of Volume II, Tocqueville's conclusions, with which we are already familiar to a large extent. Modern societies are torn by two revolutions. One tends to achieve the growing equality of conditions, the uniformity of standards of living, but also to concentrate the administration at the top, to reinforce the powers of administrative management; while another revolution is constantly weakening the traditional powers.

Given these two revolutions, revolt against traditional power, on the one hand, and administrative centralization, on the other, democratic societies are faced with the alternative with which we are so familiar: free institutions or despotism. "Thus in our days two contrary revolutions appear to be going on, the one continually weakening the supreme power, the other as continually strengthening it; at no other period in our history has it appeared so weak or so strong."

The antithesis is tempting, but it is not accurately stated. What Tocqueville meant is that power is weakened and its sphere of influence is widened. What he really had in mind is the widening of administrative and governmental functions and the weakening of the political power of decision. But the antithesis would perhaps have been less rhetorical and less striking if Tocqueville had contrasted widening, on the one hand, and weakening, on the other, instead of contrasting strengthening and weakening as he did.

These, very rapidly summarized, are Tocqueville's central ideas in the second volume of *Democracy in America,* together with some examples of his sociological method.

To conclude, I should like to attempt to characterize Alexis de Tocqueville as a political thinker and as a sociologist.

As a political thinker, Tocqueville, as he said himself, was a solitary. He derived from the Legitimist Party; his family had been attached to the French monarchy, and it was not without hesitation or mental scruples that Tocqueville supported the Orleans dynasty. In a sense, he had broken with his family tradition, and he was anxious about betraying a family he deeply loved. Nevertheless, he had attached to the Revolution of 1830 the hope that his political ideal was at last to be realized, i.e., the democratization of society combined with the strengthening of the liberal institutions in that synthesis which seemed contemptible to Auguste Comte and ideal to himself: a constitutional monarchy.

The Revolution of 1848, on the other hand, dismayed him, for he saw it as the proof, definitive for the moment, that French society was incapable of political liberty.

Thus he stood alone, neither with the Legitimists nor with the Orleanists. In parliament, he had belonged to the dynastic opposition, while explaining to the dynastic opposition that by trying to obtain a reform of the electoral law by demogogic propaganda measures, it would overthrow the dynasty. Late in January, 1848, in answer to the address to the throne, he delivered a prophetic speech in which he foretold the coming revolution. In all honesty, I should add that Tocqueville, while writing his recollections after the Revolution of 1848, admitted quite frankly that he was a better prophet than he had thought at the time he delivered this speech. For, he said, "I predicted the Revolution, my listeners thought I was exaggerating, and so did I." The Revolution broke out about a month after the moment he predicted it, amid a general skepticism which he himself shared.[24]

Politically, then, Tocqueville belonged to that liberal

party which probably had little chance of finding even a disputatious satisfaction in the course of French politics.

Tocqueville the sociologist belonged to the lineage of Montesquieu. He combined the method of the sociological portraitist with the habit of classifying types of regimes and types of societies, and with the propensity to construct abstract theories from a small number of facts. He differed from so-called classic sociologists like Comte and Marx in his rejection of vast syntheses intended to forecast history. He did not believe that past history is governed by inexorable laws or that future events are predetermined. Tocqueville, like Montesquieu, wanted to make history intelligible, but he did not want to do away with it. And in the last analysis, sociologists of the Comtian and Marxian type are always inclined to do away with history; for when we seek to know history in advance, we deprive it of its own dimension, which is action; and, when we say action, we are also saying unpredictability.

BIOGRAPHICAL CHRONOLOGY

1805   July 29. Birth in Verneuil of Alexis de Tocqueville, the third son of Hervé de Tocqueville and Madame Hervé de Tocqueville, herself born Rosambo, the granddaughter of Malesherbes, who was not only the former director of the Librairie at the time of the Encyclopédie but also lawyer to Louis XVI. Tocqueville's father and mother had been prisoners in Paris during the Reign of Terror and were saved from the guillotine by the 9-Thermidor. During the Restoration his father was prefect in several Departments, including Moselle and Seine-et-Oise.

1810–25 Studied under Father Lesueur, his father's former private tutor. Attended the *lycée* in Metz. Studied law in Paris.

1826–27 Traveled in Italy with his brother Edouard. Stayed in Sicily.

1827   Appointed by royal ordinance *juge-auditeur* in the court at Versailles, where his father had lived for a year as prefect.

1828   Met Mary Motley. Became engaged.

1830   Tocqueville unwillingly swore allegiance to Louis-Philippe. He wrote his fiancée: "I have just taken an oath. I don't reproach myself, but I am deeply distressed nonetheless and will count this day among the most unhappy of my life."

1831   Tocqueville and Gustave de Beaumont, his friend, requested and obtained from the Minister of the Interior a mission to study the American penitentiary system in the United States.

1831–32   From May 1831 to February 1832, he traveled through America, as far north as New England and Quebec, South to New Orleans, and West to Lake Michigan.

1832   Tocqueville resigned his post as magistrate out of loyalty to Beaumont, who was recalled for having refused to give evidence in an affair in which he felt the public minister had not been honorable.

1833   *Du système penitentiaire aux Etats-Unis et de son application en France,* with an appendix on the colonies by MM. G. de Beaumont and A. de Tocqueville, lawyers in the Royal Court of Paris and Members of the Historical Society of Pennsylvania.

            Traveled in England where he met Nassau William Senior.

1835   Publication of Volumes I and II of *Democracy in America* which met with immediate success.

            Traveled again in England and Ireland.

1836   Married Mary Motley.

            Wrote an article for the *London and Westminster Review* on "The Social and Political Situation in France Before and After 1789."

            Traveled in Switzerland from mid-July to mid-September.

1837   Tocqueville ran for the legislature for the first time; having refused official support, despite an offer from his relative Count Mole, he lost.

1838   Elected member of the Academy of Moral and Political Science.

1839   Tocqueville was elected by a great majority Deputy of Vologne, the district where the Tocqueville's château stood. From this time on until he retired from politics in 1851, he was continually re-elected in this district. He became the leading spokesman in proposing a law to abolish slavery in the colonies.

1840   Chairman of a bill for prison reform. Publication of

Volumes III and IV of *Democracy in America*. The reception was more reserved than in 1835.

1841 Tocqueville was elected to the *Académie française*. Trip to Algeria.

1842 Elected Counselor General of Manche, representing Sainte-Mère-Eglise and Montebourg.

1842–44 Member of the Extra-parliamentary Commission on African Affairs.

1846 October to December. Another trip to Algeria.

1847 Reporter on the extraordinary loans allotted to Algeria. In his report Tocqueville took a stand on the Algerian question. He advocated a firm attitude toward the poverty-stricken Muslims, but expressed concern for their well-being and asked that the government encourage European colonization as much as possible.

1848 January 27. Delivered a speech in the House: "I believe we are sleeping on a volcano."

April 23. In the elections to the Constituent Assembly Tocqueville maintained his position.

June. Member of the committee in charge of preparing the new constitution.

December. Tocqueville voted for Cavaignac in the presidential elections.

1849 June 2. Tocqueville became Minister of Foreign Affairs. He chose Arthur de Gobineau as his private secretary and appointed Beaumont ambassador to Vienna.

October 30. Tocqueville was forced to resign. (For this period of his life, one should read his *Souvenirs*.)

1850–51 Tocqueville edited his *Souvenirs*.

After December 2 he retired from public life.

1853 Living near Tours, he systematically researched in the city's archives the wealth of documents dating from before 1789, in order to acquaint himself with society in the Old Regime.

1854 June to September. Traveled in Germany to study the feudal system and its remains.

1856 Publication of the First Part of *The Old Regime and the Revolution*.

1857 Trip to England to consult some documents on the history of the Revolution. For his return to France, the British Admiralty put at his disposal a warship as a tribute.

1859 Died in Cannes on April 16.

NOTES

1. "The nation, taken as a whole, will be less brilliant, less glorious, and perhaps less strong; but the majority of the citizens will enjoy a greater degree of prosperity, and the people will remain peaceable, not because they despair of a change for the better, but because they are conscious that they are well off already." (Tocqueville, *Democracy in America*, Alfred A. Knopf, 1945, Vol. 1, p. 10.)

2. Tocqueville writes: "It is evident to all alike that a great democratic revolution is going on among us; but all do not look at it in the same light. To some it appears to be novel but accidental, and, as such, they hope it may still be checked; to others it seems irresistible, because it is the most uniform, the most ancient, and the most permanent tendency that is to be found in history." (*Op. cit.*, p. 3.) "The gradual development of the principle of equality is, therefore, a providential fact. It has all the chief characteristics of such a fact: it is universal, it is lasting, it constantly eludes all human interference, and all events as well as all men contribute to its progress. . . . The whole book that is here offered to the public has been written under the influence of a kind of religious awe produced in the author's mind by the view of that irresistible revolution which has advanced for centuries in spite of every obstacle and which is still advancing in the midst of the ruins it has caused. . . . If the men of our time should be convinced, by attentive observation and sincere reflection, that the gradual and progressive development of social equality is at once the past and the future of their history, this discovery alone would confer upon the change the sacred character of a divine decree. To attempt to check democracy would be in that case to resist the will of God; and the nations would then be constrained to make the best of the social lot awarded to them by Providence." (*Ibid.*, pp. 6–7.)

3. Tocqueville wrote pages on America's superiority with respect to industry, and fully recognized the major characteristic of American society. See especially chapters XVIII, XIX, and XX of the second book of the second volume of *Democracy in America*. Chapter XVIII is entitled, "Why among the Americans all Honest Callings are Considered Honorable"; Chapter XIX, "What Causes Almost All Americans to Follow Industrial Callings"; and Chapter XX, "How an Aristocracy May Be Created by Manufactures."

In Chapter XIX, Tocqueville writes, "The Americans arrived

but as yesterday on the territory which they inhabit, and they have already changed the whole order of nature for their own advantage. They have joined the Hudson to the Mississippi, and made the Atlantic Ocean communicate with the Gulf of Mexico, across a continent of more than five hundred leagues in extent which separates the two seas. The longest railroads that have been constructed up to the present time are in America." (*Democracy in America*, Vol. 2, pp. 156–57.)

4. In Chapter XX of *Democracy in America*, "How an Aristocracy May Be Created by Manufactures," Tocqueville writes, "In proportion as the mass of the nation turns to democracy, that particular class which is engaged in manufactures becomes more aristocratic. Men grow more alike in the one, more different in the other; and inequality increases in the less numerous class, in the same ratio in which it decreases in the community. Hence it would appear, on searching to the bottom, that aristocracy should naturally spring out of the bosom of democracy." Tocqueville bases this observation upon an analysis of the psychological and social effects of division of labor. The worker who spends his life making the heads of pins—Tocqueville borrows the example from Adam Smith—is dehumanized. He becomes a good worker only by being less of a man, less of a citizen—here one thinks of certain passages in Marx. The master, on the other hand, acquires the habit of command, and in the vast world of business, his mind arrives at an intelligence of wholes. And this at the same moment that industry is drawing to it the rich and cultivated men of the former ruling classes. Tocqueville hastens to add, however, "But this kind of aristocracy by no means resembles those kinds which preceded it . . ." The conclusion is very characteristic of the method and sentiments of Tocqueville: "I am of opinion, on the whole, that the manufacturing aristocracy which is growing up under our eyes is one of the harshest which ever existed in the world; but at the same time it is one of the most confined and least dangerous. Nevertheless, the friends of democracy should keep their eyes anxiously fixed in this direction; for if ever a permanent inequality of conditions and aristocracy again penetrate into the world, it may be predicted that this is the gate by which they will enter." (*Democracy in America*, Vol. 2, pp. 160–61.)

5. Cf. G. W. Pierson, *Tocqueville and Beaumont in America*, New York, Oxford University Press, 1938; Doubleday Anchor Books, 1959.

The second part of the first volume of the French edition of Tocqueville's complete works published by Gallimard contains a long annotated bibliography on the problems discussed in

Democracy in America. This bibliography was prepared by Jakob Peter Mayer.

6. See Leo Strauss, *On Tyranny,* Political Science Classics, New York, 1948; *Natural Right and History* (ref.); *Persecution and the Art of Writing,* Glencoe, The Free Press, 1952; *The Political Philosophy of Hobbes: its Basis and its Genesis,* Chicago, University of Chicago Press, 1952.

According to Leo Strauss, "Classical political science took its bearings by man's perfection or by how men ought to live, and it culminated in the description of the best political order. Such an order was meant to be one whose realization was possible without a miraculous or non-miraculous change in human nature, but its realization was not considered probable, because it was thought to depend on chance. Machiavelli attacks this view both by demanding that one should take one's bearings, not by how men ought to live, but by how they actually live, and by suggesting that chance could or should be controlled. It is this attack which laid the foundation for all specifically modern political thought." (*On Tyranny,* p. 95.)

7. One thinks of J. F. Gravier's *Paris et le désert français* (first edition, Paris, Le Portulan, 1947; second edition, completely revised, Paris, Flammarion, 1958). Indeed, the first chapter of this book bears as its legend a quotation from *The Old Regime and the French Revolution.* By the same author see also *L'Aménagement du territoire et l'avenir des régions françaises,* Paris, Flammarion, 1964.

8. Chapter 4 of Part III of *The Old Regime and the French Revolution* is entitled, "How, though the Reign of Louis XVI was the most prosperous period of the monarchy, this very prosperity hastened the outbreak of the Revolution." (*The Old Regime and the French Revolution,* Doubleday Anchor Books, New York, 1955, p. 169.) This idea, which was relatively new in his time, has been adopted by modern historians of the Revolution. A. Mathiez writes, "It was not in an exhausted country but on the contrary in a flourishing, highly developed country that the Revolution was to break out. Poverty, which sometimes brings on riots, cannot provoke the great social upheavals; these always arise from an imbalance among the classes." (*La Révolution française,* Vol. 1, *La Chute de la Royauté,* Paris, Armand Colin, 1951 [first edition in 1921], p. 13.) The idea is clarified and developed by Ernest Labrousse in his great work, *La Crise de l'économie française à la fin de l'Ancien Régime et au début de la Révolution,* Paris, Presses Universitaires Françaises, 1944.

9. Between 1890 and 1913 in Russia the number of industrial workers increased from one and a half million to three

million. Production from industrial enterprises quadrupled. Production of coal increased from 5.3 to 29 million tons, that of steel from .7 to 4 million tons, and that of petroleum from 3.2 to 9 million tons. With due allowance for fluctuation in the value of currency according to Prokopowicz, the national income increased as a whole by 40% and per capita by 17% between 1900 and 1913. Advances in education were comparable. In 1874 only 21.4% of the people knew how to read and write. In 1914 this figure had changed to 67.8%. Between 1880 and 1914 the number of pupils in elementary schools increased from 1,141,000 to 8,147,000. Indeed, as early as 1899 Lenin remarked in *Capitalism in Russia* that advances in industry were more rapid in Russia than in western Europe and even in America. He added, "The development of capitalism in young countries is very much accelerated by the aid and example of old countries." In 1914 a French economist, Edmond Théry, after returning from a long study trip in Russia, wrote in a book entitled *La Transformation économique de la Russie,* "If things in the great European nations happen between 1912 and 1950 the way they have between 1900 and 1912, by the middle of the century Russia will dominate Europe from the political as well as the economic and financial standpoint." The characteristics of Russian growth before 1914 were: 1) the very considerable participation of foreign capital (which expressed itself on the level of exchanges by a considerable deficit in the commercial balance); 2) the very modern and very concentrated structure of capitalism; and 3) the strong influence of the czarist state in the formation of the infrastructure as well as in the organization of financial channels of circulation.

10. Hippolyte Taine, *Les Origines de la France contémporaine,* Paris, Hachette, 1876–93. Taine's book consists of three parts: I. *L'Ancien Régime* (two volumes); II. *La Révolution* (six volumes); and III. *Le Régime moderne* (three volumes). The discussion of the role of the intellectuals in the crisis of the Old Regime and the course of the Revolution is found in Volumes 3 and 4 of the first part. These books are entitled, *L'esprit et la doctrine* and *La propagation de la doctrine.* See especially chapters 2 (*L'esprit classique*), 3, and 4 of Volume 3.

11. Daniel Mornet, *Les Origines intellectuelles de la Révolution,* Paris, 1933.

12. *The Old Regime and the French Revolution,* pp. 148 ff. Chapter 2 of Part III is entitled, "How vehement and widespread anti-religious feeling had become in eighteenth-century France and its Influence on the nature of the Revolution."

13. "By and large—and despite the all too obvious shortcomings of some of its members—there has probably never been

a clergy more praiseworthy than that of Catholic France just before the Revolution; more enlightened, more patriotic, less wrapped up in merely private virtues, more concerned with the public good, and, last but not least, more loyal to the Faith—as persecution clearly proved in the event. When I began my study of the old regime I was full of prejudices against our clergy; when I ended it, full of respect for them." (*The Old Regime and the French Revolution,* p. 114.)

14. This portrait begins with these words: "When I observe France from this angle I find the nation itself far more remarkable than any of the events in its long history. It hardly seems possible that there can ever have existed any other people . . ." (*The Old Regime and the French Revolution,* p. 210.) Tocqueville prefaces it with these words: ". . . without a clear idea of the old regime, its laws, its vices, its prejudices, its shortcomings, and its greatness, it is impossible to comprehend the history of the sixty years following its fall. Yet even this is not enough; we need also to understand and bear in mind the peculiarities of the French temperament."

15. Tocqueville is quite aware of this difficulty. In the Preface which opens the second volume of *Democracy in America,* he writes: "But I must warn the reader immediately against an error that would be very prejudicial to me. Because I attribute so many different effects to the principle of equality, it might be inferred that I consider this principle as the only cause of everything that takes place in our day. This would be attributing to me a very narrow view of things.

"A multitude of the opinions, sentiments, and instincts that belong to our times owe their origin to circumstances that have nothing to do with the principle of equality or are even hostile to it. Thus, taking the United States for example, I could easily prove that the nature of the country, the origin of its inhabitants, the religion of the early settlers, their acquired knowledge, their previous habits, have exercised, and still do exercise, independently of democracy, an immense influence upon their modes of thought and feeling. Other causes, equally independent of the principle of equality, would be found in Europe and would explain much of what is passing there.

"I recognize the existence and the efficiency of all these various causes; but my subject does not lead me to speak of them. I have not undertaken to point out the origin and nature of all our inclinations and all our ideas; I have only endeavored to show how far both of them are affected by the equality of men's conditions." (*Democracy in America,* Vol. 2, pp. xi–xii.)

16. Book One, Chapter VIII: "How Equality Suggests to the

Americans the Idea of the Infinite Perfectability of Man." (*Op. cit.*, pp. 33–34.)

17. Book One, Chapter X: "Why Americans Are More Addicted to Practical than to Theoretical Science." (*Op. cit.*, pp. 41–47.)

18. Book One, Chapters XIII to XIX, especially Chapter XIII: "Literary Characteristics of Democratic Times," and Chapter XVII, "Of Some Sources of Poetry among Democratic Nations."

19. Book One, Chapter XX: "Some Characteristics of Historians in Democratic Times." (*Op. cit.*, pp. 85–88.)

20. Raymond Aron, *Dix-huit leçons sur la société industrielle*, Paris, Gallimard, *Idées* series, 1962; *La Lutte de classes*, Paris, Gallimard, *Idées* series, 1964.

21. "In democratic communities the majority of the people do not clearly see what they have to gain by a revolution but they continually and in a thousand ways feel what they might lose." (*Democracy in America*, Vol. 2, p. 253.)

"If America ever undergoes great revolutions, they will be brought about by the presence of the black race on the soil of the United States; that is to say, they will owe their origin, not to the equality, but to the inequality of condition." (*Ibid.*, p. 256.)

22. "I remember as if it were yesterday a certain evening in a château where my father was living at the time and to which a family celebration had brought a large number of our close relatives. The servants had been dismissed. The whole family was gathered around the hearth. My mother, who had a sweet and affecting voice, began to sing a famous air born of our civil disturbances, whose words described the misfortunes and death of King Louis XVI. When she stopped everyone wept; not for the many personal miseries they had suffered, or even for the many relatives they had lost in the civil war and on the scaffold, but over the fate of this man who had died over fifteen years before and whom the greater part of those who were shedding tears over him had never seen. But this man had been the king." (Quoted by J. P. Mayer in *Alexis de Tocqueville*, Paris, Gallimard, 1948, p. 15.)

23. Tocqueville concludes this chapter as follows, "In every democratic army the non-commissioned officers will be the worst representatives of the peaceful desires of the country, and the private soldiers will be the best. The latter will carry with them into military life the strength or weakness of the manners of the nation; they will display a faithful reflection of the community. If that community is ignorant and weak, they will allow themselves to be drawn by their leaders into disturbances, ei-

ther unconsciously or against their will: if it is enlightened and energetic, the community will itself keep them within the bounds of order." (*Democracy in America,* Vol. 2, p. 274.)

24. In the J. P. Mayer edition of the *Oeuvres complètes,* this speech is found in the appendix to the second volume of *La Démocratie en Amerique* (*Oeuvres complètes,* Part I, second volume, pp. 368–69). In it Tocqueville denounced the unworthiness of the ruling class as it had been revealed by the numerous scandals at the end of Louis-Philippe's reign. And he concluded, "Do you not feel, by a kind of instinctive intuition which cannot be analyzed but which is sure, that the earth trembles again in Europe? Do you not smell—how shall I say—a wind of revolution in the air? Nobody knows where this wind comes from nor whom it is taking with it: and it is in times like these that you remain calm in the presence of the degradation of public morals—for this word is not too strong."

After the Revolution of 1848 he lived under the Republic, which he wanted to be liberal, and was Minister of Foreign Affairs for several months. His staff was headed by Arthur de Gobineau, with whom he remained close friends in spite of the radical incompatibility of their respective ideas. But at this time Gobineau was still a young man, and Tocqueville was already a celebrity. The two volumes of *Democracy in America* had appeared in 1848, but Gobineau had written neither his *Essai sur l'inégalité des races humaines* nor his great literary works (*Les Pléiades, Les Nouvelles asiatiques, La Renaissance, Adélaide* and *Mademoiselle Irnois*).

The Sociologists
and the Revolution of 1848

I

THE ATTITUDES adopted by the sociologists I have discussed toward the Revolution of 1848 seems interesting to me from several points of view.

First of all, the Revolution of 1848, the brief duration of the Second Republic, and the *coup d'état* of Napoleon III represented, successively, the destruction of a constitutional monarchy in favor of a republic and the destruction of the republic in favor of an imperial, authoritarian regime; and in the background of all these events, there was always the threat or specter of a socialist or quasi-socialist revolution. This period from 1848 to 1851 saw the temporary domination of a provisional government in which the socialist influence was strong; next, the struggle between the Constituent Assembly and the people of Paris; next, the struggle of a Legislative Assembly (with a monarchist majority which supported the Republic) against a president elected by universal suffrage who wanted to re-establish an authoritarian empire.

In other words, in the course of the period from 1848 to 1851, France experienced a political conflict which, more than any other episode in the history of the nineteenth century, resembles the political conflicts of the twentieth century. As a matter of fact, in this period one can observe a triangular conflict between what are known in the twentieth century as fascists, more or less liberal democrats, and socialists, which we find again between 1920 and 1933 in

Weimar, Germany, and which is still observable to a certain extent in present-day France.

The French socialists of 1848 were not exactly like the communists of today; the Bonapartists of 1850 were neither Mussolini's Fascists nor Hitler's National Socialists. But it is true, nonetheless, that this period of French political history in the nineteenth century already presented the principal opponents and rivalries typical of the twentieth century.

Moreover, this intrinsically interesting period has been discussed, analyzed, and criticized by the three sociologists I have examined: Comte, Marx, and Tocqueville. The judgments they passed on its events are characteristic of their doctrines and help us to understand the importance of the abstract theories I have been expounding.

The case of Auguste Comte is the easiest. He was, quite simply, overjoyed at the destruction of those representative and liberal institutions which he regarded as linked to critical, metaphysical, and therefore anarchistic spirit and also to a blind worship of the peculiarities of the political evolution of Great Britain. In his youthful *Opuscules,* Comte had made a comparison between the political evolutions of France and England. In England, he wrote, the aristocracy had allied itself with the House of Commons, the bourgeoisie, and even with the people, thus gradually reducing the monarchy's influence and authority. France's political evolution had been the opposite; the monarchy had allied itself with the communes and the bourgeoisie, thus reducing the aristocracy's influence and authority. Now, in Comte's eyes, the English parliamentary system was merely the form assumed by an aristocracy's domination. The British parliament was the institution through which the aristocracy ruled in England.

Thus, to Comte, the parliamentary system was not a political institution of universal vocation but a mere accident of English history. To seek to introduce in France representative institutions copied from England would be, therefore, a fundamental historical error, since the essential conditions for the parliamentary system were not present. To seek to combine parliament and monarchy was, in Comte's eyes, to commit a radical error, since the French Revolution must be the enemy of the monarchy in which the old regime found its highest expression. The combination of monarchy and parliament which was the ideal of the Constituent Assembly seemed impossible to Comte, since it rested on an error of principle regarding the nature of political institutions in general and the history of France in particular.

I should add that Comte favored centralization, which he found consistent with the law of French history. He went so far as to regard the distinction between laws and decrees as a pointless refinement of legal metaphysicians and, because of this interpretation of history, to be delighted at the abolition of the French parliament in favor of what he called a temporal dictatorship. Comte was gratified that Napoleon III did away once and for all with parliamentary windbags, or what Karl Marx would have called "parliamentary cretinism."[1]

The passage most characteristic of Comte's political and historical thinking on this subject is in Volume VI of the *Cours de Philosophie Positive.*

> According to our historical theory, by virtue of the prior concentration of the various elements of the old regime around the royalty, it is clear that the primary effort of the French Revolution irrevocably to renounce the former organization must necessarily consist in a direct conflict between popular might and royal power, whose preponderance alone had characterized such a system since the end of the second modern phase. But, although this preliminary period could not indeed have had any other political aim than to effect the imminent elimination of the royalty, which the boldest innovators would not at first have dared to imagine, it is remarkable that the constitutional metaphysic was, on the contrary, then envisioning the indissoluble union of the monarchical principle with popular ascendency, and of the Catholic constitution

with the emancipation of the mind. Such incoherent speculations would no longer merit the slightest philosophical attention were it not necessary to see them as the first direct testimony of a general aberration which still exerts the most deplorable influence towards radically disguising the true nature of modern reorganization, by reducing its fundamental regeneration to a feeble universal imitation of the transitory constitution peculiar to England.

I might add parenthetically that what Auguste Comte called a "general aberration" persists in the middle of the twentieth century, since the "transitory constitution peculiar to England," i.e., representative institutions, tends to spread with varying degrees of success, to be sure, throughout the entire world. If Comte were still alive, he would find that this aberration was becoming increasingly aberrant.

To continue the quotation:

> Such, indeed, was the political utopia of the principal leaders of the Constituent Assembly, and they assuredly pursued its direct realization insofar as was possible at the time, in view of its radical contradiction of all the tendencies characteristic of French sociability.
>
> This, then, is the natural place to apply our historical theory directly to the rapid development of this dangerous illusion. Although too crude in itself to demand any special analysis, the gravity of its consequences obliges me to point out to the reader the principal foundations of this view, which he will moreover be able to develop for himself without difficulty, from the explanations to be found in the two preceding chapters.

And then the last sentence:

> The absence of any sound political philosophy makes it easy to imagine what empirical temptation must have determined such an aberration, which surely must have been quite inevitable, since it was able utterly to seduce the reason of the great Montesquieu.

A little further on, Comte established an analogy between the political regimes of England and of Venice. In his eyes, the English parliamentary constitution did not differ in essence from the Venetian constitution which was fully determined by the end of the fourteenth century. Let me quote a passage.

It may in fact be observed that the Venetian regime, fully characterized by the end of the fourteenth century, in all respects assuredly constitutes the political system most analogous to the English government as a whole, considered in the definitive form which it was to assume three centuries later. This necessary similarity evidently results from a parallel fundamental tendency, social progression toward a temporal dictatorship of the aristocratic element. It is even incontestable that, owing to the differences in times, the Venetian type must have been much more complete than the English, since it assured the ruling aristocracy a much more marked preponderance over both the central power and the popular strength. The only essential difference between the comparative destinies of these two equally transitory regimes of which the second, having arisen at a later period in the political decomposition, certainly could not pretend to the same total duration as the first, consists in the fact that Venetian independence was bound to disappear under the necessary decadence of its particular government, while English nationality must, happily, remain altogether intact amid the inevitable collapse of its provisional constitution.

Whatever the value of such an analogy, which seemed convenient in describing my historical appreciation of the English system, barring, moreover, any idea of actual imitation, it remains incontestable that, in spite of empty metaphysical theories concocted after the fact as to the imaginary weight of the various powers [this is a condemnation of Montesquieu's idea of the balance of powers], the automatic preponderance of the aristocratic element must have provided, in England as in Venice, the universal principle of a political mechanism whose real movement would surely be incompatible with this fantastic balance.

To this fundamental condition for a like regime, must be added two others which are very important, and even more peculiar to England, and which have contributed greatly to the preservation of this exceptional system, in spite of the universal active tendency toward the radical decomposition of the old order whose particular existence it is primarily destined to prolong. The first consists in the institution of Anglican Protestantism which assured the permanent subordination of the spiritual power much better than the species of Catholicism peculiar to Venice could have done. As for the second complementary condition, it concerns the spirit of political isolation eminently peculiar to England.

Thus the English constitution, like the Venetian, rests on the power of the aristocracy. Were it not for this power of the aristocracy, a constitution based on the balance of powers would result in a paralysis of government, and society would be in danger of collapse. Moreover, England has had the advantage of two exceptional conditions: the Anglican Church, which has meant the subordination of spiritual power to temporal power, and her insular situation, which has permitted a sort of solidarity among the social classes and by the same token favored the reign of the aristocracy.

As a consequence of this general theory, Comte was not displeased to see a temporal dictator put an end to the futile imitation of English institutions and to the apparent reign of the windbag metaphysicians of parliament; and in the *Système de Politique Positive* he expressed this satisfaction. He even went so far as to write, in the introduction to Volume II of the *Système,* a letter to the Czar, a temporal dictator whom he still described as "empirical," in which he expressed the hope that this empirical dictator might be instructed by the master of positive philosophy and thus make a decisive contribution to the fundamental reorganization of European society.

This dedication to the Czar provoked several different movements within the positivist school; and I should add

that in Volume III the tone was somewhat changed, owing
to a temporary aberration on the part of the temporal dictator. I refer to the Crimean War, for Comte had declared
the age of great wars to be historically at an end, and he
seems to have assigned responsibility for the Crimean War
to the Czar. He congratulated the temporal dictator of
France for having nobly ended the temporary aberration of
the temporal dictator of Russia.

This attitude toward parliamentary institutions is not, if
I may speak in the style of Auguste Comte, exclusively
related to the individual peculiarities of the great master of
positivism. This habit of condemning parliamentary institutions as "metaphysical" or "British" still exists today. I
receive rather regularly a slim publication called *Nouveau
Régime*, originally inspired by the positivist method, which
contrasts the "fiction" of parties and parliament with the
"real France." The contributors to this review are, in fact,
very intelligent. They are in search of another mode of
representation than the one we are familiar with in parties
and parliament. Comte did not want to do away with representation altogether, but he thought it sufficient for an assembly to convene every three years to approve the budget.

Here, then, is a basic attitude characteristic of Auguste
Comte and related to the primary inspiration of sociology.
For sociology, as Comte conceived it and as Durkheim
practiced it, took as its center the social and not the political. But the subordination of the political to the social may
find expression in the depreciation of parliament in favor
of "the fundamental social reality." Durkheim shared the
indifference, tinged with hostility or contempt, that the
coiner of the word "sociology" had for parliamentary institutions. Durkheim was passionately interested in social
matters, in problems of professional ethics, in the reorganization of professional organizations; what went on in parliament seemed to him secondary, if not contemptible.

A change of climate, a change of author: let us turn to
Tocqueville.

We know from Tocqueville's general attitude that he regarded as the great project of the Revolution precisely what Auguste Comte described as an aberration. As a matter of fact, Tocqueville regretted the failure of the Constituent Assembly, that is, the failure of the bourgeois reformists who wanted to combine representative institutions with the monarchy. We know further that he considered important, if not essential, that administrative decentralization which Comte regarded with the sovereign contempt of the sociologist. We also know that Tocqueville was concerned with those metaphysical schemes of the constitutionalists which Comte dismissed in a few lines as unworthy of serious investigation.

Let us now turn to the more specific matter of the Revolution of 1848.

First of all, we must not forget that Tocqueville's social position was altogether different from Comte's. The latter was an examiner at the *École Polytechnique;* he lived for a long time on the small salary he received for this work; he lost this position; then he lived on the subsidies of his positivist followers. A solitary thinker, he created a religion of humanity of which he was both prophet and high priest. This was a peculiar situation which necessarily gave an extreme form to the expression of his ideas, one not directly suited to the complexities of events.

At the same period of 1848, Alexis de Tocqueville, descendant of an old family of the French aristocracy, was deputy from La Manche to the Legislative Assembly under the reign of Louis Philippe. When the Revolution of 1848 broke out, he was in Paris. Unlike Auguste Comte, he left his apartment. He walked through the streets; he experienced certain emotions at the sight of what was happening around him. Then, when the elections to the Constituent Assembly came around, he returned to his department (La Manche) and was elected by an overwhelming majority. Thus, he was a member of the Constituent Assembly and played a rather important role, since he belonged to the committee that drew up the constitution of the Second Republic.

Finally, in May, 1849, when Louis Bonaparte (who was to become Napoleon III) was president of the Republic, Tocqueville was admitted as minister of foreign affairs to the reconstituted cabinet of Odilon Barrot. He was minister of foreign affairs for five months, until early November, 1849, when the president of the Republic, Louis Bonaparte, repealed this ministry, which was already revealing too parliamentary a nature and was dominated by the old dynastic opposition to Louis Philippe, i.e., by the liberal monarchist party, become republican out of resignation, since the restoration of the monarchy was temporarily impossible.

Therefore, at this period Tocqueville was a monarchist turned conservative republican because there was no possibility of restoring the Legitimist or Orleanist monarchy; but at the same time he was hostile to what he called the "bastard monarchy," whose imminent danger he foresaw. The bastard monarchy was the empire of Louis Bonaparte, which all observers of any perspicacity foresaw from the day the overwhelming majority of the French people voted, not for Cavaignac, that republican general and defender of the bourgeois order, but for Louis Bonaparte, who had scarcely anything in his favor but his name, his uncle's prestige, and a few ridiculous escapades. This, then, was Tocqueville's situation. What was his reaction?

Tocqueville's reaction on the day of the Revolution, February 24, was near despair, despondency. We know why. Tocqueville was a liberal conservative, resigned to democratic modernity, and passionately interested in intellectual, personal, and political freedom. For him, these freedoms were embodied in representative institutions, and he considered revolutions unfavorable to the preservation of liberty; he was convinced that revolutions, as they gather momentum, render the survival of freedom increasingly improbable.

Tocqueville's reactions to these events may be found in an absolutely thrilling book called *The Recollections of Alexis de Tocqueville*. It is the only book Tocqueville ever wrote *au courant de la plume*. Tocqueville worked and reworked his books, endlessly pondering and correcting them. But in the case of the events of 1848, he dashed off his recollections for his own personal satisfaction, and he expressed himself with an admirable frankness, since he had forbidden their publication. He expressed opinions devoid of indulgence with regard to several of his contemporaries —opinions which provide invaluable testimony as to the real feelings of the actors in the great drama as well as in the wings of history.

First of all, here is a passage in which he summed up his reactions to the revolutions he had witnessed.

> On July 30, 1830, at daybreak, I had met the carriages of King Charles X on the outer boulevards of Versailles, with damaged escutcheons, proceeding at a foot pace, in Indian file, like a funeral, and I was unable to restrain my tears at the sight. This time [in 1848] my impressions were of another kind, but even keener. Both revolutions had afflicted me; but how much more bitter were the impressions caused by the last! I had until the end felt a remnant of hereditary affection for Charles X; but that King fell for having violated rights that were dear to me, and I had every hope that my country's freedom would be revived rather than extinguished by his fall. But now this freedom seemed dead; the Princes who were fleeing were nothing to me, but I felt that the cause I had at heart was lost.
>
> I had spent the best days of my youth amid a society which seemed to increase in greatness and prosperity as it increased in liberty; I had conceived the idea of a balanced, regulated liberty, held in check by religion, custom and law; the attractions of this liberty had touched me; it had become the passion of my life; I felt that I could never be consoled for its loss, and that I must renounce all hope of its recovery.

In another passage, he tells how, returning home one evening, he met Ampère, a fellow academician. Ampère, Tocqueville tells us, was a typical man of letters; he was delighted with the Revolution, which seemed to him to correspond to his ideal. Partisans of reform outweighed reactionaries like Guizot; the monarchy was collapsing, and Ampère saw on the horizon vistas of prosperity and a republic. Ampère and Tocqueville, according to the latter's report, had a very heated argument on the subject of the day: was the Revolution a happy or an unhappy event?

It goes without saying that Tocqueville, writing several years later, was more convinced than ever that the Revolution of 1848 was an unhappy event. From his point of view he could not judge otherwise, since the end result of the Revolution of 1848 was to replace a semilegitimate, liberal, and moderate monarchy with what Auguste Comte called a temporal dictator, with what Tocqueville called a bastard monarchy, and with what is more commonly known as an authoritarian empire. It is difficult to believe that, from a political point of view, the regime of Louis Bonaparte was superior to that of Louis Philippe and his successor; but it goes without saying that here we are dealing with judgments colored by personal preferences. Even today, in French history textbooks, the enthusiasm of Ampère is better represented than the gloomy skepticism of Tocqueville. The two characteristic attitudes of the French intelligentsia—i.e., revolutionary enthusiasm, whatever the consequences, and skepticism as to the final result of such upheavals—these two attitudes persist today, and will probably still persist in the next generation.

Tocqueville tried to explain the causes of the Revolution. The most characteristic passage is in his usual manner: the February Revolution, like all great events of its kind, is born of general causes, completed, as it were, by accidents, and it would be just as superficial to derive it necessarily from the former as to attribute it exclusively to the latter.

This view is characteristic of Tocqueville's historical sociology: neither inexorable necessity nor a series of accidents. There are general causes, but these general causes are still not sufficient to account for a particular event which might have been otherwise, had such and such a secondary incident been different.

Here is Tocqueville's analysis of the general causes.

The industrial revolution which, during the past thirty years, had turned Paris into the principal manufacturing city of France and attracted within its walls an entire new population of workmen (to whom the works of the fortifications had added another population of laborers at present deprived of work) tended more and more to inflame this multitude. Add to this the democratic disease of envy, which was silently permeating it; the economical and political theories which were beginning to make their way and which strove to prove that human misery was the work of laws and not of Providence, and that poverty could be suppressed by changing the conditions of society. . . .

I should like to interpose a comment. Here we see the deeply conservative liberal who thinks that social inequalities are part of the eternal order of Providence or that at least for his age, in his time, these inequalities are impossible to extirpate.

. . . The contempt into which the governing class, and especially the men who led it, had fallen, a contempt so general and so profound that it paralyzed the resistance even of those who were most interested in maintaining the power that was being overthrown; . . . .

Another comment. Contempt for the men who govern in France is an endemic phenomenon which occurs at the end of every regime and explains the relatively bloodless character of many French revolutions, since as a rule the regime falls when no one any longer wants to fight for it. We saw a phenomenon of this kind, still more developed, not long ago, 110 years after 1848. The political classes which govern France fall from time to time into a contempt so general that it paralyzes the very people who have most reason to defend themselves.

. . . The centralization which reduced the whole revolutionary movement to the overmastering of Paris and the seizing of the machinery of government; and lastly, the mobility of all things, institutions, ideas, men and customs, in a fluctuating state of society which had, in less than sixty years undergone the shock of seven great revolutions, without numbering a multitude of smaller, secondary upheavals. These were the general causes without which the Revolution of February would have been impossible.

The principal accidents which led to it were the passions of the dynastic opposition, which brought about a riot in proposing a reform; the suppression of this riot, first over-violent, and then abandoned; the sudden disappearance of the old Ministry, unexpectedly snapping the threads of power, which the new ministers, in their confusion, were unable either to seize upon or to reunite; . . . .

Again we see the analogy. It happens that a regime in France falls during a change of ministry. This is what happened to the regime of Louis Philippe, as to the Fourth Republic.

. . . The mistakes and disorders of mind of these ministers, so powerless to re-establish that which they had been strong enough to overthrow; the vacillation of the generals; the absence of the only Princes who possessed either personal energy or popularity; and above all, the senile imbecility of King Louis-Philippe, his weakness, which no one could have foreseen, and which still remains almost incredible, after the event has proved it.

This is the sort of analytical, historical description of the Revolution which is characteristic of a sociologist who believes neither in the inexorable determinism of history nor in a series of accidents. Remember the phrase I used in connection with Montesquieu: that Montesquieu wanted to make history intelligible; so did Tocqueville. To make history intelligible is not to show that nothing could have happened otherwise but to discover the combination of general and secondary causes that weave, as it were, the web of history.

As for the Revolution of 1848, Tocqueville understood that initially it presented a socialist character; but since he was socially conservative at the same time that he was politically liberal, he judged very severely the socialist members of the Provisional Government who, he believed (agreeing with Karl Marx in this), had exceeded the tolerable bounds of stupidity. A little like Karl Marx also, Tocqueville, simply as an observer, noted that in a first phase, between February, 1848, and the meeting of the Constituent Assembly in May, the socialists enjoyed considerable influence in Paris and thus throughout France. But what did they do with this influence? Enough to terrify the bourgeoisie and the majority of the peasantry, but not enough to secure themselves any position of power.

In fact—and this is yet another point on which Tocqueville agrees with Marx—the socialist leaders of the Revolution of 1848 did not know how to exploit the favorable circumstances they enjoyed between February and May. Beginning with the meeting of the Constituent Assembly, they did not know whether they wanted to play a game of the Revolution or that of the constitutional regime. And at the decisive moment, they abandoned their troops, for during those horrible days of June the workers of Paris fought alone, without leaders.

Tocqueville was violently hostile both to the socialist leaders and to the June rioters. But his hostility did not blind him. On the one hand, he recognized the extraordinary courage exhibited by the Parisian workers in fighting against the regular army, and then he went on to say that the disrepute into which the socialist leaders had fallen was not, perhaps, definitive.

We know that for Marx, the Revolution of 1848 proved that henceforth the essential problem of European societies would be the social problem. He saw the revolutions of the nineteenth century as no longer political, but social. Tocqueville, who was obsessed with a concern for freedom, or what he called freedom, regarded these riots, insurrections, or revolutions as catastrophic. But he was aware of the fact that they presented a certain socialist character, and he added that if, for the moment, a socialist revolution seemed to him out of the question, if it was hard for him to visualize a regime established on foundations other than the principle of private ownership, he concluded prudently:

Will Socialism remain buried in the disdain with which the Socialists of 1848 are so justly covered? I put the question without making any reply. I do not doubt that the laws concerning the constitution of our modern society will in the long run undergo modifications: they have already done so in many of their principal parts. But will they ever be destroyed and replaced by others? It seems to me to be impracticable. I say no more, because—the more I study the former condition of the world and see the world of our own day in greater detail, the more I consider the prodigious variety to be met with not only in laws, but in the principles of law, and the different forms even now taken and retained, whatever one may say, by the rights of property on this earth—the more I am tempted to believe that what we call necessary institutions are often no more than institutions to which we have grown accustomed, and that in matters of social constitution the field of possibilities is much more extensive than men living in their various societies are ready to imagine.

After the description of the days of June, the balance of Tocqueville's *Recollections* is devoted to an account of the drawing up of the constitution of the Second Republic, his participation in the second Odilon Barrot ministry, and the struggle of liberal monarchists-turned-republicans against both the royalist majority in the Assembly and the President, whom they suspected of trying to re-establish the empire. Among the most brilliant pieces of bravura, I recommend a portrait of Lamartine of which the essential phrase is, "I do not know that I have ever, in this world of selfishness and ambition in which I lived, met a mind so void of any thought of the public welfare as his." And, of course, there is the superb portrait of Louis Bonaparte.

In conclusion, then, Tocqueville did understand the socialist character of the Revolution of 1848, but he condemned the behavior of the socialists as senseless. He belonged to the bourgeois party of order and, during the days of June, was prepared to fight against the insurgent workers. In the second phase of the crisis, he was a moderate republican, a supporter of what was later called the conservative republic; and at the same time he was an anti-Bonapartist. He was defeated, but he was not surprised at his defeat, for from the first day of the Revolution of 1848 he believed that free institutions were temporarily doomed; he thought that the Revolution of 1848 would inevitably result in an authoritarian regime of one kind or another, and after the election of Louis Bonaparte he naturally foresaw the restoration of the empire. But since it is not necessary to hope in order to act, he fought against the solution that seemed to him at once most probable and least desirable—which is characteristic of a sociologist of the school of Montesquieu.

II

MARX SPENT the period from 1848 to 1851 differently from either Comte or Tocqueville. He had not retired to the ivory tower of the Rue Monsieur le Prince, nor was he a deputy to the Constituent Assembly or the Legislative Assembly or a minister to Odilon Barrot or to Louis Bonaparte. As a revolutionary agitator, he participated in the events in Germany, but he had lived in France for many years and was well acquainted with French politics and revolutionaries. He was, therefore, an intelligent witness of the events in France; he believed in the international character of the Revolution and felt directly involved in the French crisis.

Many of the judgments to be found in his two books, *The Class Struggle in France, 1848–1850* and *The Eighteenth Brumaire of Louis-Bonaparte,* coincide with those to be found in Tocqueville's *Recollections.*

Like Tocqueville, he was struck by the contrast between the riots of 1848, in which the working masses of Paris fought alone and without leaders for several days, and the disturbances of 1849 in which, only a year later, the parliamentary leaders of La Montagne vainly tried to start riots and were abandoned by their troops. The contrast between the troops without leaders on the first date and the leaders without troops on the second was emphasized by both authors.

They were equally aware of the fact that the events of 1848–1851 no longer represented merely political disturbances but were the premonitory symptoms of a social revolution. Tocqueville observed with alarm that henceforth what was at stake was the whole foundation of society, laws respected for centuries by all human societies. Marx exclaimed triumphantly that the social upheaval which he regarded as necessary was taking place. The liberal aristocrat's scale of values was different from—in a sense, opposite to—the revolutionary's. That respect for political freedom which for Tocqueville constitutes a sacred value was for Marx the superstition of a man of the old regime. Marx had no respect for parliament, parliamentarians, formal liberties. What Tocqueville desired to save above all else, Marx believed to be secondary, perhaps even an obstacle on the path to what seems to him essential, namely, the social revolution.

Both writers saw a sort of historical logic in the transition from the Revolution of 1789 to the Revolution of 1848. Tocqueville explained this continuity by the formula: the Revolution continues. The extension of the Revolution involved questioning the social order and property ownership, after the destruction of the monarchy and the privileged orders. In this social revolution Marx discerned the rise of the fourth estate after the victory of the third. The language is not the same; once again, the value judgments are opposite; but it would not be difficult, by comparing texts, to see that on this essential point the two thinkers agreed. Once the traditional monarchy has been destroyed, once the aristocracy of the past has been overthrown, the democratic movement, striving toward social equality, must attack any privileges that still exist—and this meant the privileges of the bourgeoisie. Economic inequality is the object of attack after inequality of orders or estates has been erased.

To Tocqueville's mind, the struggle against economic inequality was, at least in his own day, doomed to failure. More often than not he seems to have regarded economic inequalities—inequalities of fortune—as inseparable from the eternal order of human societies. Marx, on the other hand, thought that through a reorganization of society these economic inequalities could be reduced or eliminated.

Finally, Marx and Tocqueville also coincided in their analysis of the phases of the Revolution of 1848–1851. The events in France during that period fascinated contemporary observers and are still absorbing today precisely because in a few years France experienced most of the situations typical of political conflicts in modern societies.

The first phase extended from February 24, 1848, through May 4, 1848, when an insurrection overthrew the monarchy and the Provisional Government included several socialists who, for a few months, exerted a dominant influence.

A second phase began with the meeting of the Constituent Assembly. The majority of the Assembly elected by the whole country was conservative or even reactionary and monarchist. Whence a first conflict between a predominantly socialist Provisional Government and a conservative Assembly, resulting in the riots of June, 1848, and the revolt of the Parisian proletariat against an Assembly which was elected by universal suffrage but seemed to the Parisian workers to be their enemy.

The third phase began with the election of Louis Bonaparte in December, 1848, or, according to Marx, with the end of the Constituent Assembly in May, 1849. The President of the Republic believed in the Bonapartist legitimacy —believed himself to be the man of destiny. As President of the Second Republic, he clashed first with a Constituent Assembly which had a monarchist majority and next with a Legislative Assembly which included a monarchist majority but also one hundred and fifty representatives of the left, La Montagne.

With the election of Louis Bonaparte, there began a subtle, many-sided conflict in which the monarchists, incapable of reaching an agreement on the identity of the monarch or on the restoration of the monarchy itself, became, through their hostility to Louis Bonaparte, the defenders of the Republic against a Bonaparte seeking to restore a bastard monarchy. Louis Bonaparte used to a certain extent the devices which parliamentarians call those of popular demagogy. There were, in Louis Bonaparte's strategy, elements of the pseudo-socialism of the Fascists and National Socialists of the twentieth century. Since the Legislative Assembly made the mistake of abolishing universal suffrage, in the *coup d'état* of the Eighteenth Brumaire Louis Bonaparte violated the constitution, dissolved the Legislative Assembly, and at the same time re-established universal suffrage.

Bearing in mind the points of agreement between Tocqueville and Marx, what are the specific characteristics of the Marxist analysis? The central idea, I think, is this: Marx sought to explain political events in terms of what might be called social infrastructure. He sought to demonstrate, in strictly political conflicts, the expression or outcropping on the political level, as it were, of the underlying conflicts of social groups. Tocqueville, quite obviously, did the same; he, too, showed social groups in conflict in the France of the mid-nineteenth century. But although he explained political conflicts in terms of social conflicts, he maintained the specificity, at least the relative autonomy, of the political order. In the France of 1848, the principal actors in the drama were the peasants, the Parisian *petite bourgeoisie*, the Parisian workers, and the remnants of the aristocracy. The principal actors in the drama as seen by Tocqueville were not very different from the principal actors as seen by Marx. (I have quoted Tocqueville's statement that only classes concern the historian.) But Marx always tried to find an exact correspondence between events on the political level and events in the social infrastructure. What concerns us here, with regard to a sociological theory, is to determine to what extent he succeeded.

Let me insert a parenthetical comment. Marx's two pamphlets to which I refer, *The Class Struggle in France* and *The Eighteenth Brumaire of Louis-Bonaparte,* are brilliant works; in many respects I think they are more profound and more satisfying than his massive scientific books. Marx, concentrating on historical insight, forgot his theories and analyzed events as an observer—and as an observer of genius.

To give a first example of his interpretation of politics in terms of social infrastructure, I shall quote a passage from *The Class Struggle in France.*

> December 10, 1848 [i.e., the day of Louis Bona-parte's election], was the day of the peasant insurrection. Only from this day does the February of the French peasants date. The symbol that expressed their entry into the revolutionary movement, clumsily cunning, knavishly naïve, doltishly sublime, a calculated superstition, a pathetic burlesque, a cleverly stupid anachronism, a world historical piece of buffoonery and an undecipherable hieroglyphic for the understanding of the civilized—this symbol bore the unmistakable features of the class that represents barbarism within civilization. The republic had announced itself to this class with the tax collector; it announced itself to the republic with the emperor. Napoleon was the only man who had exhaustively represented the interests and the imagination of the peasant class, newly created in 1789. By writing his name on the front page of the republic, it declared war abroad and the enforcing of its class interests at home. Napoleon, for the peasants, was not a person but a program. With banners, with beat of drums and blare of trumpets,

they marched to the polling booths shouting: *plus d'impôts, à bas les riches, à bas la république, vive l'Empereur*. No more taxes, down with the rich, down with the republic, long live the emperor! Behind the emperor was hidden the peasant war. The republic that they voted down was the republic of the rich.

This is an admirable piece of rhetoric, but what does it mean? The sentence that follows gives us the clue: "December 10 was the *coup d'état* of the peasants which overthrew the existing government."

What are the possible interpretations of Louis Bonaparte's election? The commonplace, traditional interpretation is something like this: The French peasants, who represented a majority of the electors at the time, chose to elect the real or supposed nephew of the Emperor Napoleon rather than the republican general, Cavaignac. According to a psychopolitical interpretation, Louis Bonaparte, because of his name, was the charismatic leader. Max Weber would certainly have written that the peasant (that uncivilized creature, according to Marx, with his contempt for peasants) had preferred a Napoleonic symbol to a genuine republican personality. In this sense, Louis Bonaparte was the man of the peasantry against the republic of the rich.

Obviously, even the non-Marxist admits that the peasants elected Louis Bonaparte. What seems problematical is the extent to which Louis Bonaparte, by virtue of the fact that he was elected by the peasants, became the representative of their class interest. The peasants were under no necessity to choose Louis Bonaparte to interpret their class interest, nor were the measures taken by him necessarily consistent with the peasants' class interest. The emperor acted as his genius or his stupidity inspired him to act.

In other words, the incontestable event is the peasant vote for Louis Bonaparte. The transformation of the event into theory is the proposition: "The peasants' class interest was represented by Louis Bonaparte."

Another passage pertaining to the peasants deserves to

be quoted. It is from *The Eighteenth Brumaire of Louis-Bonaparte;* Marx was describing the class situation of the peasants.

> The peasants are an immense mass whose individual members live in identical conditions, without however entering into manifold relations with one another. Their method of production isolates them from one another, instead of drawing them into mutual intercourse. This isolation is promoted by the poor means of communication in France and by the poverty of the peasants themselves. Their field of production, the small allotment of land that each cultivates, allows no room for a division of labor, and no opportunity for the application of science; in other words, it shuts out manifoldness of development, diversity of talent, and the luxury of social relations.
>
> Every single peasant family is almost self-sufficient; it itself directly produces the greater part of what it consumes; and it earns its livelihood more by means of an interchange with nature than by intercourse with society. We have the allotted patch of land, the peasant and his family; alongside of them another allotted patch of land, another peasant and another family. A group of these makes up a village; a group of villages makes up a Department. Thus the large mass of the French nation is constituted by the simple addition of equal magnitudes, much as potatoes in a bag constitute a potato bag.
>
> Insofar as millions of families live under economic conditions that separate their mode of life, their interests and their culture from those of other classes, and that place them in a hostile attitude toward the latter, they constitute a class. Insofar as there exists among these peasants only a local connection in which the individuality and exclusiveness of their interests prevent any unity of interest, national connections and political organization among them, they do not constitute a class.
>
> Consequently, they are unable to assert their class interests in their own name, whether through a parliament or through a convention. They cannot represent

themselves, they must be represented. Their representative must at the same time appear as their master, as an authority over them, as an unlimited governmental power that protects them from the other classes, and that bestows rain and sunshine upon them from above.

This is a very penetrating description of the ambiguous position—a class, yet not a class—of the peasant masses. The peasants had a more or less similar mode of existence, which gave them the primary characteristic of a social class; but they lacked the capacity to become conscious of themselves as a unit. They were also incapable of representing themselves. They were, as it were, a passive class that could be represented only by outsiders, a situation which begins to explain the fact that the peasants chose a man who was not a peasant, Louis Bonaparte.

The difficulty with such a system of interpretation is this: Is what happens on the political level adequately explained by what happens in the social infrastructure?

For example, according to Marx, the Legitimist monarchy represented landed property; the Orleanist monarchy represented the industrial and commercial bourgeoisie. The two dynasties were never able to reach an agreement. During the crisis of 1848–1851, the insurmountable obstacle to the restoration of the monarchy was, precisely, the quarrel between the two dynasties. The problem of interpretation is this: Were the two royal families incapable of agreeing on the identity of the pretender because one represented landed property and the other industrial and commercial property or because, by definition, there can be only one pretender?

The question is not inspired by a predisposition to criticism or overrefinement; it raises the essential problem in interpreting a political situation in terms of social infrastructure. Let us admit that Marx was right: that the legitimist monarchy actually is the regime of the great landowners and the traditional nobility. Let us agree, further, that the Orleans monarchy did represent the interests of the industrial and commercial bourgeoisie. Was it a conflict of economic interests that prevented unity or the simple arithmetic fact, so to speak, that there can be only one authentic pretender and not two?

Marx, naturally, was inclined to explain the impossibility of agreement in terms of an incompatibility of economic interests.[2] The weakness of this interpretation is that there have been other countries and other circumstances in which the interests of landed property could effect a compromise with the interests of the industrial and commercial bourgeoisie.

I shall quote the passage which is most interesting on this point.

> The diplomats of the party of order [you may recall that "the party of order" referred at the time to the majority of the Constituent Assembly, then of the Legislative Assembly, composed mainly of monarchists, some of whom were legitimists, others Orleanists] believed they could allay the struggle by a combination of the two dynasties through a so-called fusion of the royalist parties and their respective royal houses. The true fusion of the restoration and the July Monarchy was, however, the parliamentary republic, in which the Orleanist and Legitimist colors were dissolved and the various species of bourgeois vanished in the plain bourgeois, in the bourgeois genus.

Here we have a purely political interpretation which is accurate and satisfying. The two monarchist parties could reach an agreement only with a parliamentary republic, which is the only way to reconcile two pretenders to a throne that permits only one occupant. There is no mystery here. When there are two pretenders to the throne, to avoid having one in the Louvre and the other in prison or exile, neither must be in power. In this sense, the parliamentary republic was a reconciliation between the two dynasties.

Marx continued: "Now, however, the plan was to turn the Orleanist into a Legitimist and the Legitimist into an Orleanist." Marx was right in thinking that such a thing is out of the question, unless by a stroke of good luck the pretender of one of the two families should be good enough to die.

The monarchy, in which their antagonism was personified, was to incarnate their unity, the expression of their exclusive factional interest was to become the expression of their common class interest, the monarchy was to accomplish what only the abolition of two monarchies—the Republic—could do and had done.

This was the philosophers' stone, for the finding of which the doctors of the party of order were breaking their heads. As though the Legitimate monarchy could ever be the monarchy of the hereditary landed aristocracy! As though landed property and industry could fraternize under one crown when the crown could fall only upon one head, the head of the older or the younger brother! As though industry could deal upon an equal footing with landed property so long as landed property did not itself decide to become industrial!

We see, in these last two sentences, a subtle and persuasive mingling of two explanations: the political explanation, whereby the only way to reconcile the two opposed pretenders to the French throne was the parliamentary republic, and the other, fundamentally different interpretation, whereby landed property could not be reconciled with the industrial bourgeoisie unless landed property itself was industrialized. This explanation belongs, I may say, to the category of the more intellectual explanations still found in the Marxist reviews of today regarding the Fifth Republic. The latter cannot be the Gaullist republic; it must be the republic either of modernized capitalism or of some other element of the social infrastructure.[3] Such explanations do, to be sure, go deep, but they have the disadvantage, in my opinion, of not being true. For the impossibility of reconciling the interests of landed property with the interests of the industrial bourgeoisie exists only in the sociological imagination. On the day when, by a stroke of good luck, one of the two princes no longer has any descendants, the reconciliation between the two pretenders occurs automatically and the interests of landed property and industrial property miraculously effect a compromise. In other words, the impossibility of reconciling the two pretenders is essentially political.

The explanation of many political events in terms of their social foundation is valid, but the term-for-term explanation of what happens on the political level in terms of what supposedly happens within the social infrastructure is to a large extent sociological mythology. We project onto the social infrastructure what we have observed on the political level. Since we have observed that the two pretenders cannot agree, we decree that landed property cannot agree with industrial property. Moreover, we contradict ourselves when we explain that this reconciliation is achieved in a parliamentary republic. For if such a reconciliation were *socially* impossible, it would be just as impossible under a parliamentary republic as under a monarchy.

I have analyzed the case at some length because, in my opinion, it is typical. It demonstrates both what is acceptable, instructive, and necessary in social explanations of political conflicts and what is erroneous. Professional and amateur sociologists suffer from a sort of guilty conscience when they confine themselves to explaining changes of regime and political crises in terms of politics. Personally, however, I am inclined to believe that the particulars of political events can rarely be explained except in terms of men and parties, their disagreements and their ideas.

Was Louis Bonaparte the representative of the peasants? Yes, he was elected by the peasant electors. Is General de Gaulle the representative of the peasants? He was elected by 85 percent of the French people. The psychopolitical mechanism was the same a century ago as it is today. It has little to do with distinctions of social class and still less with the class interest of a given group. When the French people are tired of endless conflicts and there appears a candidate for the national salvation, a man of destiny,

Frenchmen of all classes rally to the man who promises to save them and who sometimes actually does so.

I should like to take a last example, a last case which concerns one of our constant problems, namely, the relation between economics and politics.

In the last part of *The Eighteenth Brumaire of Louis-Bonaparte,* Marx made an extremely interesting and detailed analysis, which unfortunately I do not have space to follow in detail, of Louis Bonaparte's government and the way in which he served the interests of the different classes. Marx's essential idea was that Louis Bonaparte was accepted by the bourgeoisie because he defended their fundamental economic interests, with the bourgeoisie, for their part, relinquishing the right to exercise political power themselves. But there is an element in this analysis which interests us especially, i.e., Marx's acknowledgment of the decisive role of the state. He wrote:

> This executive power, with its tremendous bureaucratic and military organization, with its widespread and artificial machinery of government, with an army of officeholders half-a-million strong, together with a military force of another million men, this fearful body of parasites, which coils itself like a snake around French society, stopping all its pores, originated at the time of the absolute monarchy, with the decline of feudalism, which it helped to hasten. The princely privileges of the landed proprietors and cities were transformed into so many attributes of the executive power, the feudal dignitaries into paid officeholders, and the confusing design of conflicting medieval seigniories into the well-regulated plan of a government, whose work is subdivided and centralized as in a factory.
>
> The first French revolution, with its mission of sweeping away all local, territorial, urban and provincial special privileges in order to establish the civic unity of the nation, was bound to develop what the absolute monarchy had begun—the work of centralization, together with the range, the attributes and the

agents of governmental authority. Napoleon completed this governmental machinery. The Legitimists and the July Monarchy contributed nothing to it, except a greater subdivision of labor, which grew in the same measure that the division and subdivision of labor within bourgeois society raised new groups and interests, that is, new material for the administration of government.

Each common interest was in turn forthwith removed from society, set up against it as a higher collective interest.

In other words, Marx was describing the prodigious development of the centralized administrative state. This centralized administrative state is the same one which Tocqueville analyzed, whose prerevolutionary origins he demonstrated, and which he saw gradually gaining in importance and power as democracy developed.

Whoever governs this state inevitably exerts considerable influence on society. Tocqueville was convinced that all parties contribute to the growth of this enormous administrative machine. He was convinced that a socialist state would contribute even more toward broadening the functions of the state and centralizing administration. Marx believed that this machine of the state had acquired a kind of autonomy in relation to society. All it takes is "some swindler just off the boat, exalted to the skies by a group of drunken soldiers he has bought with brandy and sausages" to manage this enormous machine. Marx was a sociologist; but he was also a polemicist.

The true revolution would consist not in taking over this machine but in destroying it. To which Tocqueville would have replied ironically: If you decree that ownership of the instruments of production must become collective and that the administration of the economy must be centralized, by what miracle do you expect the machine of the centralized state to be destroyed or reduced?

As a matter of fact, there are in Marx two ideas of the role of the state in the revolution. In his remarks on the

Paris Commune, he implied that the Commune—the break-down of the central state and its decentralization—was indeed the dictatorship of the proletariat. But we also find exactly the opposite idea: political power and state centralization must be strengthened to the maximum degree in order to bring about the revolution.

On the machine of the centralized state, then, Tocqueville and Marx held similar views. Tocqueville concluded that it was essential to strengthen the intermediate bodies and representative institutions with a view to limiting the omnipotence and boundless extent of the state. Marx both acknowledged the state's partial autonomy in relation to society—a formula contradictory to his general theory of the state as a mere expression of the ruling class—and assumed more or less vaguely that the socialist revolution might entail the destruction of the centralized administrative machine.

What interests me here, and what I should like to make my concluding point, is the following idea: In theory, Marx tried to reduce politics and its conflicts to the relations between social classes and their conflicts. But on several points, his insight as an observer prevailed over his dogmatism as a theorist, and he recognized involuntarily, as it were, the strictly political factors in such conflicts and the autonomy of the state itself in relation to various groups. But to the extent that there is state autonomy in relation to classes and class conflicts, one element in the evolution of societies is not reducible to the struggle of social classes. What decisions are made depends on who takes possession of the state.

I should add that the most striking proof of the autonomy of the political order in relation to social conflicts is the Russian Revolution of 1917, when a group of men, by seizing the state as Louis Bonaparte did but more violently, was able to transform the entire structure of Russian society and even, if you will, to establish socialism, not on the basis of the predominance of a minority proletariat but on the basis of the omnipotence of the state machine.

In other words, what is not found in official Marxist theory may be found either in Marx's historical analyses or in events whose participants acted in the name of Marx himself.

A few words by way of conclusion. The four authors studied here form the basis of three movements or schools. The first is the one that might be called the French school of political sociology, whose founder is Montesquieu and whose second great figure is Tocqueville. A man like Élie Halévy, in our time, belongs to this tradition:[4] a school of sociologists who are not very dogmatic, who are essentially preoccupied with politics, who do not disregard the social infrastructure but stress the autonomy of the political order, and who are liberals. Probably I am a latter-day descendant of this school.

The second school is that of Auguste Comte, culminating in Durkheim at the beginning of this century and, if I may venture to say so, in the official and licensed sociologists of today. This is the school which underplays the political as well as the economic in relation to the social. It places the emphasis on the unity of the social entity, retains the notion of consensus as its fundamental concept, and by multiplying analyses and concepts endeavors to reconstruct the social totality.

The third school—the one that has been most successful, not in classrooms, but on the great stage of world history —is obviously the Marxist school. This school, as interpreted by hundreds of millions of human beings, combines an explanation of the social entity in terms of economic organization and social infrastructure with a schema of evolution that guarantees its followers victory and the peaceful or violent elimination of heretics.

This school of sociology is the most difficult to discuss, because of its historical success. One never knows whether one should discuss the catechist version, necessary to any state doctrine, or the intellectual version, the only one acceptable to great minds. But between the catechist version and the intellectual version there is a constant interaction whose modalities vary, obviously, according to the unpredictable vicissitudes of universal history.

These three schools of sociology, despite the differences between their systems of values and their views of world history, are all interpretations of modern society. Auguste Comte was an almost unqualified admirer of modern society, which he called industrial, because this society would be peaceful and Comtist or, if you prefer, positivist. In the eyes of the political school, modern society is a democratic one which a Tocqueville or an Élie Halévy observes, without transports of enthusiasm or of indignation, one society among others—one, to be sure, with singular characteristics, but not the ultimate fulfillment of human destiny. As for the third school, it combines a Comtist enthusiasm for industrial society with a righteous indignation against capitalism. It is, therefore, supremely optimistic about the distant future and darkly pessimistic about the near future. It predicts a long period of catastrophe, class struggles, and wars before the final reconciliation of the classes and of all human societies with one another.

In other words, the Comtist school is optimistic, with a tendency to complacency; the political school is cautious, with a tinge of skepticism; and the Marxist school is utopian, with a tendency to accept catastrophes as desirable in the long run and, in any event, inevitable.

Each of these schools reconstructs the social entity in its own way. Each offers its own theory of the diversity of societies known to history. Each is inspired both by moral convictions and by scientific hypotheses. I have tried, for whatever it is worth, to separate the convictions from the hypotheses. But, as usual, even when we attempt to distinguish these two elements, we do so, perhaps, in terms of our own convictions.

CHRONOLOGY OF EVENTS
IN THE REVOLUTION OF 1848
AND THE SECOND REPUBLIC

1847–48   Agitation for electoral reform in Paris and the provinces: the Banquet Campaign.

1848   February 22. Banquet and a reformist manifestation in Paris despite a ministerial ban.

23. Parisian National Guard demonstrated to the cry "Long Live the Reform." Guizot resigned. In the evening the people and troops clashed and corpses of demonstrators were dragged through the Parisian streets.

24. Revolution declared by morning. Republican insurgents seized the Hôtel de Ville, threatening the Tuileries. Louis-Philippe abdicated in favor of the Count of Paris, his grandson, and fled to England. Insurgents surrounded the House of Deputies to prevent the queenship from being conferred on the Duchess of Orléans. A provisional government was established by evening and included Dupont de L'Eure, Lamartine, Crémieux, Arago, Ledru-Rollin, Garnier-Pagès. The Government Secretaries were A. Marrast, Louis Blanc, Flocon, and Albert.

25. Proclamation of the Republic.

26. Abolition of the death penalty for political crimes. Creation of national workshops.

27. Titles of nobility abolished.

March 2. Settled on ten-hour day in Paris, eleven-hour day in the provinces.

5. Elections for the constitution of the Constituent Assembly.

6. Garnier-Pagès became Minister of Finance. Increased the supplementary tax on all direct contributions by forty-five centimes to the franc.

16. Demonstrations by the National Guard's bourgeois element in protest against dissolution of the Elite Companies.

17. Mass demonstrations by the supporters of the provisional government. Socialists and leftist Republicans asked that elections be postponed.

April 16. Another mass demonstration to postpone elections. Provisional government appealed to the National Guard to restrain them.

23. Nine hundred representatives elected to the Constituent Assembly. Progressive Republicans won only 80 seats, Legitimists about a hundred, and the Orléanists 200. The majority of the Assembly, that is, about 500 seats, went to the Moderate Republicans.

May 10. Assembly appointed an executive committee of five members in order to insure the government: Arago, Garnier-Pagès, Lamartine, Ledru-Rollin, Marie.

15. Barbes, Blanqui, and Raspail led demonstration in favor of Poland. Demonstrators surrounded the House of Deputies at Hôtel de Ville. Announced a new insurrectionary government to the crowd. National Guard arrested Barbes and Raspail and dispersed the demonstrators.

June 4–5. Louis-Napoleon Bonaparte was elected Deputy in three Departments of the Seine.

21. Dissolution of National workshops.

23–26. Riots. The eastern and central sections of Paris fell into the hands of insurgent Parisian workers, who were able to take shelter behind barricades because of the inaction of Cavaignac, the Minister of War.

24. Assembly granted Cavaignac full power; he crushed the uprising.

July to November. Constitution of the Party of the Order. Thiers campaigned for Louis-Napoleon Bonaparte, equally popular among the workers. Drawing up of the constitution by the National Assembly.

November 12. Proclamation of the Constitution which provided for a Chief Executive to be elected by universal suffrage.

December 10. Elected the President of the Republic. Louis-Napoleon received 5,500,000 votes, Cavaignac 1,400,000, Ledru-Rollin 375,000, and Lamartine 8000.

20. Louis-Napoleon swore allegiance to the Constitution.

1849   March to April. Trial and condemnation of Barbes, Blanqui, Raspail, the organizers of revolutionary attempts in May 1848.

April to July. Expedition of Rome. French Expeditionary Troop took the city, re-establishing Pope Pius the Ninth.

May. Elections to the Legislative Assembly which numbered 75 Moderate Republicans, 180 Montagnards and 450 Monarchists (Legitimists as well as Orléanists) from the Party of the Order.

June. Demonstrations in Paris and Lyon against the Expedition of Rome.

1850   March 15. Falloux Law which reorganized public education.

May 31. Electoral law that compelled any resident of three months in one district to vote. It prevented about three million itinerant workers from voting.

May to October. Socialist agitation in Paris and in the surrounding area.

August to September. Negotiations between the Legitimists and the Orléanists to re-establish the monarchy.

September to October. Military reviews at Camp Satory before the President-Prince. The cavalry paraded to the cry "Long Live the Emperor."

Clash between the Assembly majority and the President-Prince.

1851   July 17. General Magnan, who was devoted to the President-Prince, was appointed Military Governor of Paris in place of Changarnier, who was faithful to the Monarchist majority in the Assembly.

December 2. Coup d'état. State of siege declared. Dissolution of the Assembly; re-establishment of universal suffrage.

20. Napoleon with 7,350,000 votes against 646,-000 was elected for a ten-year term, receiving the power to write a new constitution.

1852   January 14. Enactment of the new constitution.

November 20. A new plebiscite approved, with 7,840,000 votes against 250,000, the re-establishment of the Imperial office in the person of Louis Napoleon who took the title of Napoleon the Third.

NOTES

1. Comte, however, is not of the Bonapartist tradition. After the Lycée de Montpellier he was even very hostile to the policy and legend of Napoleon. Except for the period of the Hundred Days, when Comte, then a student at the Ecole Polytechnique, was carried away by the Jacobin enthusiasm that reigned in Paris, Bonaparte was for him the model of the great man who, through failure to understand the course of history, led the country backward and would leave nothing behind him. On December 7, 1848, on the eve of the presidential election, he

wrote to his sister, "I, who have never wavered from the sentiments which you knew me to have toward the retrograde hero in 1814, would consider the political restoration of his race a disgrace to my country." Later, he would speak of the "fantastic vote of the French peasants, who might just as well confer on their fetish a life span of two centuries with immunity to the gout." But, on December 2, 1851, he applauded the coup d'état, preferring a dictatorship to the parliamentary Republic and to anarchy, an attitude which even provoked the departure of Littré and the liberal disciples from the positivist society. But this did not prevent Comte from calling a "preposterous masquerade" the combination of popular sovereignty and hereditary principle that underlay the restoration of the Empire in 1852 or from prophesying then the collapse of the regime in 1853. Auguste Comte entertained several times—in 1851, and then in 1855 when he published the Appeal to the Conservatives—the hope that Napoleon III would be converted to positivism. But he also frequently looked hopefully to the proletarians, whom he admired for their philosophical purity, which he contrasted with the metaphysics of the lettered. In February 1848, his heart was with the revolution. In June, shut up in his apartment in Rue-Monsieur-le-Prince, near the barricades that surrounded the Panthéon where very violent fighting was going on, Comte was for the proletarians against the government of metaphysicians and littérateurs. When he speaks of the insurgents he says "we," but he pities them for still being seduced by the utopias of the "reds," those "apes of the great revolution." Thus Comte's political attitude in the course of the Second Republic may appear very fluctuating and full of contradictions, but it is the very logical consequence of a thought that placed the success of positivism above everything, could not recognize it in any of the parties, and in any case saw the revolution only as a transitory anarchic crisis. However, one sentiment dominated all others: contempt for parliamentarianism.

A passage from the preface to the second volume of the *Système de politique positive,* published in 1852, on the eve of the re-establishment of the Empire, represents Auguste Comte's synthesis of the events of the past four years:

"It seems to me that our latest crisis has brought the French Republic irrevocably out of the parliamentary phase, which could only be appropriate to a negative revolution, and into the dictatorial phase, which alone is suited to the positive revolution which will bring about the gradual termination of the western malady by means of a decisive reconciliation between order and progress.

"Even if an abuse of the dictatorship which has just appeared made it necessary to change its principal organ before the appointed time, this unfortunate necessity would not truly reestablish the domination of some kind of assembly, except perhaps during the short interval required by the exceptional advent of a new dictator.

"According to the historical theory which I have originated, all of French history has always tended to promote the domination of the central power. This normal disposition would never have ceased, had not this power finally acquired a retrograde character in the second half of the reign of Louis XIV. This resulted, a century later, in the total abolition of French royalty; whence the temporary domination of the only truly popular assembly this nation has ever seen [that is, the Convention].

"Indeed its ascendency resulted only from its proper subordination to the energetic Committee which arose from its midst to direct the heroic republican defense. The need to replace the royalty with a true dictatorship soon made itself felt, due to the sterile anarchy which resulted from our first attempt at a constitutional regime.

"Unfortunately, this indispensable dictatorship soon also took a profoundly backward direction, combining the enslavement of France with the oppression of Europe.

"It was only as a reaction to this deplorable policy that French opinion then permitted the only serious attempt ever made among us at a regime peculiar to the English situation.

"It suited us so ill that, in spite of the benefits of western peace, its official preponderance for a generation became even more baneful to us than the imperial tyranny, perverting minds by the habit of constitutional sophisms, corrupting hearts by vicious or anarchic morals, and breaking down character under the growing rise of parliamentary tactics.

"Due to the fatal absence of any real social doctrine, this disastrous regime persisted in other forms after the republican explosion of 1848. This new situation, which spontaneously guaranteed progress and turned all serious concern toward order, made doubly necessary the normal ascendency of the central power.

"However, it was then believed that the elimination of a useless royalty would bring about the full triumph of the opposing force. All those who had actively participated in the constitutional regime, the government, the opposition, or the conspiracies, should, four years ago, have been irrevocably banished from the political scene as incapable or unworthy of directing our Republic.

"But a blind enthusiasm on all sides granted them the supremacy of a constitution which specifically established the parliamentary omnipotence. Universal suffrage extended to the proletarians the intellectual and moral ravages of this regime which until then had been confined to the upper and middle classes.

"Instead of the preponderance which it should have resumed, the central power, which thus lost the prestige of inviolability and perpetuity, nevertheless retained the constitutional incapacity which these honors had previously concealed.

"Reduced to such an extremity, this necessary power happily reacted with energy against an intolerable situation, which was as disastrous for us as it was shameful to itself.

"The popular instinct in no way resisted the fall of an anarchic regime. One feels increasingly in France that constitutionality is suited only to an alleged monarchical situation, whereas our republican situation permits and requires dictatorship." (*Système de politique positive,* Vol. II, Preface, Letter to M. Vieillard, February 28, 1852, pp. xxvi–xxvii.)

On all this, see H. Gouhier, *La Vie d'Auguste Comte,* second edition, Paris, Vrin, 1965, and *La Jeunesse d'Auguste Comte et la formation du positivisme,* Vol. 1, Paris, Vrin, 1933.

2. In this connection a passage from *The Eighteenth Brumaire* is significant: "Legitimists and Orléanists constituted, as we have said, the two great factions of the party of order. Was not what bound these factions to their pretenders and opposed them to each other something other than the fleur-de-lis and the tricolor, the house of Bourbon and the house of Orléans, which were merely different nuances of royalism? Under the Bourbons it was the *great landed property* that had reigned, with its priests and its lackeys. Under the Orléans, it was high finance, heavy industry, big business, that is, *capital,* with its retinue of lawyers, professors, and fine talkers. The legitimate royalty was merely the political expression of the hereditary rule of landowning lords, just as the July monarchy was merely the political expression of the usurped rule of bourgeois upstarts. Consequently, what divided the factions among themselves was not some alleged principles, but their material conditions of existence, two different kinds of property, the old antagonism between town and country, the rivalry between capital and landed property. That old memories, personal enmities, fears and hopes, prejudices and illusions, sympathies and antipathies, convictions, articles of faith, and principles also bound them to one or the other royal house, is self-evident. Out of the different forms of property, out of the conditions of social existence there rises a whole superstructure of impressions, illusions, ways of thinking and particular philosophical conceptions. The class as a whole creates them and develops them on

the foundation of these material conditions and the corresponding social relations. The individual who receives them through tradition or education may imagine that they constitute the real determining factors and groundwork of his activity. If both the Orléanists and the legitimists tried to persuade themselves and others that they were divided by their attachment to their two royal houses, later events showed that it was the divergence of their interests which prevented the union of the two dynasties. And just as in private life we distinguish between what a man says or thinks of himself and what he really is and does, in historical struggles we must distinguish even more carefully between the phraseology and pretentions of parties and their real nature and interests, between what they think they are and what they really are. Orléanists and legitimists existed in the Republic side by side, with identical pretentions. If each faction aspired against the other to the *restoration* of its *own* dynasty, this signified only that each of the *two great interests* dividing the *bourgeoisie*—landed property and capital—was trying to reestablish its own supremacy and the subordination of the other. I speak of two interests of the bourgeoisie, for great landed property, in spite of its feudal affectations and ancestral pride, had become completely middle class as a result of the development of modern society."

3. See especially the articles of Serge Mallet, which have been collected in one volume entitled *Le Gaullisme et la Gauche*, Paris, Editions du Seuil, 1965. For this sociologist the new regime is not an accident of history, "but the establishment of a political structure corresponding to the requirements of neo-capitalism." Gaullism is the political expression of modern capitalism. A comparable analysis is found in Roger Priouret, who, although not a Marxist, believes that "de Gaulle did not come to power in 1958 simply by the force of the shock of Algiers; he believed he was bringing a regime based on his historical views and he in fact adapted political life to the state of the society." ("Les institutions politiques de la France en 1970," *Bulletin S. E. D. E. I. S.*, No. 786, *Futuribles* supplement, May 1, 1961.)

4. Among the works of Élie Halévy, I mention: *La Formation du radicalisme philosophique*, Paris, Alcan, 1901–4 (3 volumes: Vol. I, *La Jeunesse de Bentham;* Vol. II, *L'Evolution de la doctrine utilitaire de 1789 à 1815;* Vol. III, *Le Radicalisme philosophique); Histoire du peuple anglais au XIXᵉ siècle*, Paris, Hachette, 6 volumes (the first four volumes cover 1815 to 1848, the last two 1895 to 1914); *L'Ere des tyrannies, études sur le socialisme et la guerre*, Paris, Gallimard, 1938; *Histoire du socialisme européen*, based on lecture notes, Paris, Gallimard, 1948.

Bibliographies

MONTESQUIEU

WORKS IN ENGLISH TRANSLATION

The Spirit of the Laws, trans. Thomas Nugent. Rev. ed., 2 vols.; New York: Hafner, 1962.

CRITICAL STUDIES

Courtney, C. P. *Montesquieu and Burke.* Oxford: Basil Blackwell, 1963.

Fletcher, F. T. H. *Montesquieu and English Politics 1750–1800.* London: E. Arnold, 1939.

Hearnshaw, Fossey John Cobb, ed. *The Social and Political Ideas of Some Great French Thinkers of the Age of Reason.* London: G. G. Harrap, 1930.

Laski, Harold Joseph. *The Rise of European Liberalism,* "An Essay in Interpretation." London: George Allen & Unwin, 1936. Ch. 3.

Levin, Lawrence Meyer. *The Political Doctrine of Montesquieu's "Esprit des lois,"* "Its Classical Background." New York: Columbia University Press, 1936.

Martin, Kingsley. *French Liberal Thought in the Eighteenth Century,* "A Study of Political Ideas from Bayle to Condorcet," ed. J. P. Mayer. Rev. ed.; London: Turnstile Press, 1954.

Shackleton, Robert. *Montesquieu, A Critical Bibliography.* London: Oxford University Press, 1961.

Spurlin, Paul Merrill. *Montesquieu in America 1760–1801.* University: Louisiana State University Press, 1940.

AUGUSTE COMTE

WORKS IN ENGLISH TRANSLATION

The Positive Philosophy, freely trans. and condensed by Harriet Martineau. 3 vols.; New York: Calvin Blanchard, 1855.
Positive Polity, trans. by Frederic Harrison, E. S. Beesley, J. H. Bridges, and others. London: 1875–1877.

CRITICAL STUDIES AND BIOGRAPHIES

Barnes, Harry Elmer. *An Introduction to the History of Sociology.* Chicago: University of Chicago Press, 1948. Ch. III.
Bridges, John Henry. *The Unity of Comte's Life and Doctrines.* London: N. Trübner, 1866.
Caird, Edward. *The Social Philosophy and Religion of Comte.* Glasgow: J. Maclehose & Sons, 1885.
Fiske, John. *Outlines of Cosmic Philosophy.* 2 vols.; London: Macmillan, 1874.
Hutton, Henry Dix. *Comte, the Man and the Founder.* London: 1891.
Marvin, Francis Sidney. *Comte, the Founder of Sociology.* London: Chapman & Hall, 1936.
Mill, John Stuart. *Auguste Comte and Positivism.* Ann Arbor: University of Michigan Press, 1961.
Vaughan, Charles Edwyn. *Studies in the History of Political Philosophy before and after Rousseau,* ed. A. G. Little. 2 vols.; New York: Russell & Russell, 1960. Vol. II, ch. 5.

KARL MARX

WORKS IN ENGLISH TRANSLATION

Capital, ed. Frederick Engels and trans. Samuel Moore & Edward Aveling. 3 vols.; Chicago: Charles H. Kerr, 1909–1918.
A Contribution to the Critique of Political Economy, trans. N. I. Stone. Chicago: Charles H. Kerr, 1904.
Economic and Philosophic Manuscripts of 1844, ed. Dirk J. Struik and trans. Martin Milligan. New York: International Publishers, 1964.
The Poverty of Philosophy. Chicago: Charles H. Kerr, 1910.
Marx-Engels, *Selected Works.* 2 vols.; Moscow: Foreign Languages Publishing House, 1955.

Bibliographies

Marx-Engels, *Selected Correspondence 1846–1895*. New York: International Publishers, 1942.

CRITICAL STUDIES AND BIOGRAPHIES

Berlin, Isaiah. *Karl Marx, His Life and Environment*. 2nd ed.; New York: Oxford University Press, 1948.

Bober, Mandell Martin. *Karl Marx's Interpretation of History*. Rev. ed.; Cambridge: Harvard University Press, 1948.

Böhm-Bawerk, Eugen von. *Karl Marx and the Close of His System*, ed. Paul M. Sweezy. New York: A. M. Kelley, 1949.

Cole, George Douglas Howard. *What Marx Really Meant*. New York: Alfred A. Knopf, 1934.

Egbert, Donald Drew, and Persons, Stow, eds. *Socialism and American Life*. 2 vols.; Princeton: Princeton University Press, 1952.

Hook, Sidney. *From Hegel to Marx*, "Studies in the Intellectual Development of Karl Marx." New York: Reynal & Hitchcock, 1936.

Hook, Sidney. *Towards the Understanding of Karl Marx*, "A Revolutionary Interpretation." New York: The John Day Co., 1933.

Hunt, Carew R. N. *Marxism, Past and Present*. London: 1954.

Kautsky, Karl. *The Economic Doctrines of Karl Marx*, trans. H. J. Stenning. London: A. C. & Black, 1925.

Laski, Harold Joseph. *Karl Marx*, "An Essay." London: The Fabian Society, 1922.

Lindsay, Alexander Dunlop. *Karl Marx's "Capital,"* "An Introductory Essay." London: Oxford University Press, 1925.

Mayo, H. B. *Democracy and Marxism*. New York: Oxford University Press, 1955.

Mehring, F. *Karl Marx*, "The Story of His Life," eds. Ruth & Heinz Norden and trans. Edward Fitzgerald. New York: Covici Friede, 1936.

Meyer, Alfred G. *Marxism*, "The Unity of Theory and Practice." Cambridge: Harvard University Press, 1954.

Popper, Karl R. *The Open Society and Its Enemies*. Rev. ed.; Princeton: Princeton University Press, 1950.

Robinson, Joan. *An Essay on Marxian Economics*. London: Macmillan, 1942.

Rühle, Otto. *Karl Marx*, "His Life and Work," trans. Eden & Cedar Paul. New York: Viking Press, 1929.

Schlesinger, Rudolf. *Marx, His Time and Ours*. London: Routledge & Kegan Paul, 1950.

Wilson, Edmund. *To the Finland Station*, "A Study in the Writing and Acting of History." Garden City, N. Y.: Doubleday, 1955. Parts II–III.

ALEXIS DE TOCQUEVILLE

WORKS IN ENGLISH TRANSLATION

Democracy in America, Henry Reeve text revised by Francis Bowen and ed. Phillips Bradley. 2 vols.; New York: Alfred A. Knopf, 1945.

The European Revolution and Correspondence with Gobineau, ed. and trans. John Lukacs. New York: Doubleday Anchor Books, 1959.

Journey to America, ed. Jakob Peter Mayer and trans. George Lawrence. New Haven: Yale University Press, 1960.

The Old Regime and the French Revolution, trans. Stuart Gilbert. Garden City: Doubleday Anchor Books, 1955.

CRITICAL STUDIES AND BIOGRAPHIES

Laski, Harold Joseph. "Alexis de Tocqueville and Democracy," in Fossey John Cobb Hearnshaw, ed., *The Social and Political Ideas of Some Representative Thinkers of the Victorian Age*. London: G. G. Harrap, 1933.

Lively, Jack. *The Social and Political Thought of Alexis de Tocqueville*. Oxford: Clarendon Press, 1962.

Mayer, Jakob Peter. *Alexis de Tocqueville*, "A Biographical Study in Political Science." New York: Harper Torchbooks, 1960.

Mayer, Jakob Peter. *Prophet of the Mass Age*, "A Study of Alexis de Tocqueville," trans. MM. Bozman & C. Hahn. London: J. M. Dent & Sons, 1939.

Pierson, George Wilson. *Tocqueville in America*, abridged by Dudley Cammett Lunt from *Tocqueville and Beaumont in America*. New York: Doubleday Anchor Books, 1959.

Pierson, George Wilson. *Tocqueville and Beaumont in America*. New York: Oxford University Press, 1938.

Salomon, Albert. "Tocqueville, Moralist and Sociologist," *Social Research*, 2(1935).

Schapiro, J. Salwyn. "Alexis de Tocqueville, Pioneer of Democratic Liberalism in France," *Political Science Quarterly* (1942).

Index